Lecture Notes in Economics and Mathematical Systems 485

Springer-Verlag Berlin Heidelberg GmbH

Andrey Garnaev

Search Games and Other Applications of Game Theory

Springer

Author
Prof. Andrey Garnaev
Saint Petersburg State University of Architecture and
Civil Engineering
Department of Computational Mathematics
2-ya Krasnoarmejskaya 4
Saint Petersburg 198005, Russia

Cataloging-in-Publication Data applied for

Die Deutsche Bibliothek - CIP-Einheitsaufnahme

Garnaev, Andrej:
Search games and other applications of game theory / Andrej Garnaev.
- Berlin ; Heidelberg ; New York ; Barcelona ; Hong Kong ; London ;
Paris ; Tokyo ; Barcelona ; Budapest : Springer, 2000
(Lecture notes in economics and mathematical systems ; 485)
ISBN 978-3-540-67195-4

ISSN 0075-8442
ISBN 978-3-540-67195-4 ISBN 978-3-642-57304-0 (eBook)
DOI 10.1007/978-3-642-57304-0

Springer-Verlag is a company in the BertelsmannSpringer publishing group.

Originally published by Springer-Verlag Berlin Heidelberg New York in 2000

Typesetting: Camera ready by author
Printed on acid-free paper SPIN: 10734512 42/3143/du-543210

Preface

This book is on applications of game theory. The title of this book is not "Game Theory and its Applications" because it does not construct a general theory for considered games. The book contains a lot of examples of application of game theory together with the background of those games considered and a list of unsolved problems. Also we consider only the game where the optimal strategies of the players are found in closed form. This book is an attempt to carry on the approach developed in nice books "Search Games" by Gal and "Geometric Games and their Applications" by Ruckle.

The first chapter of this book supplies the required definitions and theorems from game theory.

The second chapter deals with discrete search games where both players act simultaneously: the games of protection of a channel from infiltration of a submarine, the submarine versus helicopter game, the matrix search games and others.

The third chapter considers the game where the players allocate their continuous efforts. In these games players face up an alternative either not to come into contest if the cost of efforts seems too high, or come into it. In the last case the player have to decide how much resources they can afford to spend. The allocation models of search, antiballistic protection and marketing are investigated.

The forth chapter studies dynamic infiltration games where the infiltrator tries to penetrate a protected zone uncaught. We treat cases in which the guard's information about the location of the infiltrator may be nonexistent or not current (with a specific lag time.) The customs versus smuggler game where the customs tries to stop the smuggler who is attempting to ship a cargo of perishable contraband across a strait is considered. Also, an inspection game is described.

The fifth chapter deals with non-zero sum games of timing. The models of shooting contests, R&D competition, dividing a cake and other are studied. The uniqueness of Nash equilibrium is investigated. It is shown that the uniqueness depends on information the player has about his rival.

The sixth chapter considers parlour games. The cover-up, exchange and poker games are investigated.

Saint Peterburg, October 1999 *Andrey Garnaev*

Contents

1 Preliminary Results from Game Theory

1.1 Zero-sum Games

A two person zero-sum game is defined as a 3-tuple (X, Y, M) where X and Y are sets and M is real valued function defined on the Cartesian product $X \times Y$. The set X is called the set of admissible pure strategies of player 1 and set Y is called the set of admissible pure strategies of player 2. The function M is called the payoff function of player 1. Player 1 chooses a strategy x of the set X while player 2 chooses a strategy y of the set Y. The choices are done simultaneously and independently and the chosen x and y determine the playoff $M(x, y)$ to player 1 and $-M(x, y)$ to player 2. So, it is considered as if each player hands his choice to a referee who then announces (x, y) and executes the payoffs). In zero-sum game the players have antagonistic interests. The payoffs can be considered as amounts of money or utilities. The data of the game (X, Y, M) are known to both players.

A pair (x^*, y^*) of strategies is called an equilibrium (a saddle-point) if the following conditions hold

$$M(x, y^*) \leq M(x^*, y^*) \leq M(x^*, y) \quad \text{for any} \quad (x, y) \in X \times Y\,.$$

If (x^*, y^*) is an equilibrium then the strategies x^* and y^* of the players are called optimal and $M(x^*, y^*)$ is called the value of the game. It is clear that if player 1 plays x^* then he can win at least $M(x^*, y^*)$ no matter what player 2 plays. Likewise if player 2 plays y^* then he can win at most $M(x^*, y^*)$ no matter what player 1 plays.

The special case in which X and Y are finite is called a finite game or a matrix game. In this case, the function M for the game (X, Y, M) is described as a payoff matrix A whose rows are labelled by the elements of X (usually denoted as $1, \ldots, n$) and the columns by the elements of Y (denoted as $1, \ldots, m$). An example of matrix game is

$$\begin{pmatrix} 1 & -1 \\ -1 & 1 \end{pmatrix}.$$

This game can be described as follows. Two players choose tail or head. If both of them choose the same side of the coin then player 2 pays one dollar to player 1. Otherwise player 1 pays one dollar to player 2. Not every game has a saddle point. In fact the above game has no.

In view of non-existence of the value for matrix games, it is self-suggested that a player can sometimes do better by choosing his strategy randomly. For example, in the mentioned game if a player chooses head and tail with equal probabilities, then his expected payoffs will be zero independerly of behaviour of the other player. This motivates the following definition.

The mixed extension of a matrix game (X, Y, M) is the game $(\bar{X}, \bar{Y}, \bar{M})$ where

$$\bar{X} = \{x \in R^n : x_i \geq 0 \text{ for } i \in [1, n],\ \sum_{i=1}^{n} x_i = 1\},$$

$$\bar{Y} = \{y \in R^m : y_i \geq 0 \text{ for } i \in [1, m],\ \sum_{i=1}^{m} y_i = 1\}$$

and

$$\bar{M}(x, y) = \sum_{i=1}^{n} \sum_{j=1}^{m} A_{i,j} x_i y_j .$$

So, the strategy set of player in the mixed extension is the set of probability distributions on the strategy set of the original game. The elements of these sets are called mixed strategies. The payoff in the mixed extension is just the expected payoff of the player. It is clear that the extreme points of the set $\bar{X}$ or $\bar{Y}$ can be identified with the strategy sets X and Y.

Theorem 1.1.1 *(The Minmax Von-Neumann Theorem [109]). The mixed extention of a finite game has a value.*

Finding optimal strategies of the players and the value of the game in closed form is a kind of art. There are only a few interesting games where it was managed to do. For solving matrix games the following theorems can be of use.

Theorem 1.1.2 *The strategies x^* and y^* of the players are optimal and v is the value of the matrix game if and only if the following conditions hold*

$$\sum_{i=1}^{n} A_{i,j} x_i^* \leq v \quad \textit{for} \quad j \in [1, m],$$

$$\sum_{j=1}^{m} A_{i,j} y_j^* \geq v \quad \textit{for} \quad i \in [1, n].$$

Theorem 1.1.3 *The strategies x^* and y^* of the players are optimal and v is the value of the matrix game if and only if they are solutions of the followig linear programming problems:*

Problem (a) *Minimize v*

$$\sum_{i=1}^{n} A_{i,j} x_i^* \leq v \quad \text{for} \quad j \in [1, m],$$

$$\sum_{i=1}^{n} x_i^* = 1, \quad x_i^* \geq 0 \quad \text{for} \quad i \in [1, n],$$

Problem (b) *Maximize v*

$$\sum_{j=1}^{n} A_{i,j} y_j^* \geq v \quad \text{for} \quad i \in [1, n],$$

$$\sum_{j=1}^{m} y_j^* = 1, \quad y_j^* \geq 0 \quad \text{for} \quad j \in [1, m].$$

Theorem 1.1.4 *Let x^* and y^* be optimal strategies of the players and v be the value of the matrix game. If*

$$\sum_{j=1}^{m} A_{i,j} y_j^* < v$$

then $y_i^* = 0$, and if

$$\sum_{i=1}^{n} A_{i,j} x_i^* > v$$

then $x_j^* = 0$.

In a matrix game we say that row i dominates row k if

$$A_{i,j} \geq A_{k,j} \quad \text{for all} \quad j$$

and

$$A_{i,j} > A_{k,j} \quad \text{for at least one} \quad j.$$

Similarly, we say that columns j dominates columns r if

$$A_{i,j} \leq A_{i,r} \quad \text{for all} \quad i$$

and

$$A_{i,j} < A_{i,r} \quad \text{for at least one} \quad i\,.$$

Theorem 1.1.5 *A player in his optimal behavior can always dispanse without dominated strategies and use only undominated strategies.*

Theorem 1.1.6 *Let A be a $n \times n$ marix game and*

$$A_{i,j} = -A_{j,i} \quad \textit{for all} \quad i \textit{ and } j\,.$$

This game is called symmeric, and its value is zero. Optimal strategy of a player is also optimal for his rival.

1.2 Non-zero-sum Games

A two person game is defined as a 4-tuple (X, Y, M_1, M_2) where X and Y are sets and M_1 and M_2 are real valued function defined on the Cartesian product $X \times Y$. The set X is called the set of admissible pure strategies of player 1 and set Y is called the set of admissible pure strategies of player 2. The function M_i is called the payoff function of player i where $i = 1, 2$. Player 1 chooses a strategy x of the set X while player 2 chooses a strategy y of the set Y. The choices are done simultaneously and independently and the chosen x and y determine the playoff $M_1(x, y)$ to player 1 and $M_2(x, y)$ to player 2. So, it is considered as if each player hands his choice to a referee who then announces (x, y) and executes the payoffs). The payoffs can be considered as amounts of money or utilities. The data of the game (X, Y, M_1, M_2) are known to both players.

A pair (x^*, y^*) of strategies is called an equilibrium (Nash equilibrium) if the following conditions hold

$$\begin{aligned} M_1(x, y^*) &\le M_1(x^*, y^*) \quad \text{for any} \quad x \in X\,, \\ M_2(x^*, y) &\le M_2(x^*, y^*) \quad \text{for any} \quad y \in Y\,. \end{aligned}$$

If (x^*, y^*) is an equilibrium then the strategies x^* and y^* are called optimal.

Theorem 1.2.1 *([77]). The mixed extention of a finite non-zero sum gane has at least one equilibrium.*

2 Ambush Games

2.1 An Infiltration Game with a Cable

Consider the following zero-sum infiltration problem. There are two players: Guard and Infiltrator (for example, coast guard and submarine.) Guard wishes to protect a channel against penetration by Infiltrator. For this purpose Guard uses an electric cable with length k. We assume Infiltrator knows the length of cable but it is invisible to him. If Infiltrator crossing the channel touches the cable he gets detected and captured by Guard. Consider the following zero-sum game modelling this problem. Guard chooses in integer interval $[1, n]$ an interval consisting of m points where $m < n$ and Infiltrator chooses an integer point, called his point of crossing, in $[1, n]$. The payoff to Guard equals 1 if the point chosen by Infiltrator is in the interval chosen by Guard and 0 otherwise. A pure strategy of Guard is denoted by i, where i is his point of crossing and a mixed strategy of Infiltrator is the probability vector $x = (x_1, \ldots, x_n)$ saying him to choose his point of crossing at i with probability x_i. We denote an integer interval consisting of m points and its left endpoint at p by $M(p)$. Then we can denote a pure strategy y of Guard by p, where $p \in [1, n-m+1]$ and p means choosing $M(p)$. A mixed strategy of Guard is the probability vector $y = (y_1, \ldots, y_{n-m+1})$ saying him to choose interval $M(i)$ with probability y_i. Let $E(x, y)$ be the expected payoff to Guard if Infiltrator and Guard use the strategy x and y, respectively. Denote by $\Gamma(m, n)$ the considered game and let $v(m, n) = \text{val}(\Gamma(m, n))$.

Theorem 2.1.1 *The value of the game $\Gamma(m, n)$ is*

$$v(m, n) = \frac{1}{\lfloor n/m \rfloor + 1},$$

where $\lfloor s \rfloor$ is the greatest integer less than s.
An optimal mixed strategy x^* of Infiltrator is to choose with equal probability a crossing point from the set $A = \{1 + im,\ \ i \in [0, \lfloor n/m \rfloor]\}$. An optimal mixed strategy y^* of Guard is to choose with equal probability an interval from the following set of intervals $B = \{[1 + im,\ (i+1)m],\ \ i \in [0,\ \lfloor n/m \rfloor - 1], [n - m + 1,\ n]\}$.

Proof Suppose that Infiltrator employs the mixed strategy x^* and Guard plays the pure strategy y. The distance between neighbouring points of A is

m so, there is no two points from A covered by $M(y)$ simultaneously. Hence, Guard will detect Infiltrator with probability at most $\frac{1}{\lfloor n/m \rfloor + 1}$. So,

$$E(x^*, y) \leq \frac{1}{\lfloor n/m \rfloor + 1}.$$

Assume that Guard adopts the mixed strategy y^* and Infiltrator plays the pure strategy x. Then since the intervals from B cover $[1, n]$ and except two last ones they do not overlap, it follows that x intersects at least one of the possible Guard's choices. So, he will be detected with probability at least $\frac{1}{\lfloor n/m \rfloor + 1}$. Then

$$E(x, y^*) \geq \frac{1}{\lfloor n/m \rfloor + 1}.$$

This completes the proof of Theorem 2.1.1.

2.2 An Infiltration Game with Two Cables

Consider a generalization of the game $\Gamma(m, k)$ where Guard possesses two electric cables with lengths k and m respectively which cannot be cut and which will detect any crossing by Infiltrator. The two cables are allowed to overlap and we assume Infiltrator knows the lengths of cables but they are invisible to him. This situation can be described by the following zero-sum game. Guard chooses in integer interval $[1, n]$ two integer intervals consisting of k and m points where $k + m < n$ and Infiltrator chooses an integer point, called his crossing point, in $[1, n]$. The payoff to Guard equals 1 if the point chosen by Infiltrator is at least in one of the intervals chosen by Guard and 0 otherwise. A pure strategy of Guard is denoted by i, where i is his point of crossing and a mixed strategy of Guard is the probability vector $x = (x_1, \ldots, x_n)$ saying him to choose his point of crossing at i with probability x_i. We denote an integer interval consisting of m (k) points and its left endpoint at $p(\xi)$ by $M(p)$ ($K(\xi)$). Then we can denote a pure strategy y of Red by (p, ξ), where $p \in [1, n - m + 1]$ and $\xi \in [1, n - k + 1]$ and $p(\xi)$ means choosing $M(p)$ ($K(\xi)$). Let $E(x, y)$ be the expected payoff to Guard if Infiltrator and Guard use the strategies x and y respectively. Denote by $\Gamma(m, k, n)$ the considered game and let $v(m, k, n) = \text{val}(\Gamma(m, k, n))$. Without loss of generality we can consider that $k \leq m$.

Theorem 2.2.1 *If $m \geq n/2$ then*

$$v(m, k, n) = \frac{\lfloor (n - m)/k \rfloor + 2}{2(\lfloor (n - m)/k \rfloor + 1)}.$$

An optimal mixed strategy x^* of Infiltrator is to choose with equal probability a crossing point from the set $A = \{1 + m + ik,\ 1 + ik,\ i \in [0, \lfloor (n - m)/k \rfloor]\}$.

An optimal mixed strategy y^* of Guard is to choose with equal probability an interval from the following set of intervals $B = \{(M(p_i), K(\xi_i)), i \in [1, 2(\lfloor \frac{n-m}{k} \rfloor + 1)]\}$, where

$$p_i = \begin{cases} 1 & \text{for } i \in [1, \lfloor \frac{n-m}{k} \rfloor + 1], \\ n - m + 1 & \text{for } i \in [\lfloor \frac{n-m}{k} \rfloor + 2, 2(\lfloor \frac{n-m}{k} \rfloor + 1)], \end{cases}$$

$$\xi_i = \begin{cases} 1 + m + (i-1)k & \text{for } i \in [1, \lfloor \frac{n-m}{k} \rfloor], \\ n - k + 1 & \text{for } i = \lfloor \frac{n-m}{k} \rfloor + 1, \\ 1 + k(i - \lfloor \frac{n-m}{k} \rfloor - 2) & \text{for } i \in [\lfloor \frac{n-m}{k} \rfloor + 2, 2(\lfloor \frac{n-m}{k} \rfloor + 1)]. \end{cases}$$

Proof Suppose that Infiltrator employs the mixed strategy x^* and Guard plays the pure strategy y. It is clear that A consists of $2(\lfloor \frac{n-m}{k} \rfloor + 1)$ points and, since $m \geq n/2$, any couple of intervals $M(m)$ and $K(k)$ intersects with the set A at most at $\lfloor \frac{n-m}{k} \rfloor + 2$ points. So,

$$E(x^*, y) \leq \frac{\lfloor (n-m)/k \rfloor + 2}{2(\lfloor (n-m)/k \rfloor + 1)}.$$

Assume that Guard adopts the mixed strategy y^* and Infiltrator plays the pure strategy x. Since $m \geq n/2$, then

- Each point $t \in [m+1, n-m]$ is covered $2(\lfloor \frac{n-m}{k} \rfloor + 1)$ times by $\{M(p_i)\}$
- Each point $t \in [1, m] \cup [n-m+1, n]$ is covered $\lfloor \frac{n-m}{k} \rfloor + 1$ times by intervals $\{M(p_i)\}$, and one time by intervals $\{K(\xi_i)\}$.

So, each point of $[1, n]$ is covered at least $\lfloor \frac{n-m}{k} \rfloor + 1$ times by couples of intervals from B. Then

$$E(x, y^*) \geq \frac{\lfloor (n-m)/k \rfloor + 2}{2(\lfloor (n-m)/k \rfloor + 1)}.$$

This completes the proof of Theorem 2.2.1.

Analogously to the proof of Theorem 2.1.1 we can prove the following two theorems.

Theorem 2.2.2 *If $m = k$ then*

$$v(m, m, n) = \frac{2}{\lfloor n/m \rfloor + 1}.$$

An optimal mixed strategy x^* of Infiltrator is to choose with equal probability a crossing point from the set $A = \{1 + im, \ i \in [0, \lfloor n/m \rfloor]\}$. An optimal mixed strategy y^* of Guard is to choose with equal probability any combination of two intervals not depending on order from the following set of intervals $\{[1 + im, (i+1)m], \ i \in [0, \lfloor n/m \rfloor - 1], [n - m + 1, n]\}$.

Theorem 2.2.3 *If $n = (m+k)r$, where r is an integer, then*

$$v(m,k,n) = (m+k)/n = 1/r\,.$$

An optimal mixed strategy x^* of Infiltrator is to choose with equal probability a crossing point from the interval $[1,n]$. An optimal mixed strategy y^* of Guard is to choose with equal probability an interval from the following set of intervals $\{(M(p_i), K(\xi_i)), i \in [1,r]\}$, where

$$\begin{aligned} p_i &= 1 + (i-1)(m+k), \\ \xi_i &= 1 + (i-1)(m+k) + m \quad \textit{for } i \in [1,r]\,. \end{aligned}$$

Assumption 2.2.1 *By Theorems from 2.2.1 to 2.2.3, we can restrict our attention to the case where $m \in (k, n/2)$ and $n \neq 0 \bmod (m+k)$.*

First consider the case $k = 1$.

Definition 2.2.1 *Let α be a non-negative integer such that $n = \alpha(m+1)+\beta$ where $\beta \in [1,\ m]$.*

Consider separately the following two cases:

$$\alpha + \beta \geq m \text{ and } \alpha + \beta < m\,.$$

Remark 2.2.1 *If $\alpha + \beta \geq m$ then*

$$n - m(m+1-\beta) = (m+1)(\alpha+\beta-m) \geq 0\,.$$

Theorem 2.2.4 *If $\alpha + \beta \geq m$ then*

$$v(m,1,n) = \frac{m+1}{n}\,.$$

An optimal mixed strategy x^* of Infiltrator is to choose with equal probability a crossing point from the interval $[1,n]$. An optimal mixed strategy y^* of Guard is to choose with equal probability an interval from the following set of intervals $B = \{(M(p_i), K(\xi_i)), i \in [1,n]\}$, where

$$\begin{aligned} p_i &= 1 + m\left\lfloor \frac{i}{m+1} \right\rfloor \quad \textit{for} \quad i \in [1,n]\,, \\ \xi_i &= \begin{cases} 1 + \alpha m + \left\lfloor \dfrac{i}{m+1-\beta} \right\rfloor & \textit{for } i \in [1, m(m+1-\beta)]\,, \\ 1 + (\alpha+1)m \\ \quad + \left\lfloor \dfrac{i - m(m+1-\beta)}{m+1} \right\rfloor & \textit{for } i \in [m(m+1-\beta)+1, n]\,. \end{cases} \end{aligned} \tag{2.1}$$

Proof By Remark 2.2.1, (2.1) correctly defines pure strategies of Guard. Assume that Guard adopts the mixed strategy y^* and Infiltrator plays the pure strategy x. It is easy to see that

- Each point $t \in [1, \alpha m]$ is covered $m+1$ times by $\{M(p_i)\}$
- Each point $t \in [1+\alpha m, (\alpha+1)m]$ is covered $n - \alpha(m+1) = \beta$ times by $\{M(p_i)\}$
- Each point $t \in [1+\alpha m, (1+\alpha)m]$ is covered $m+1-\beta$ times by $\{K(\xi_i)\}$
- By Remark 2.2.1 each point $t \in [1+(1+\alpha)m, n]$ is covered $m+1$ times by $\{K(\xi_i)\}$.

So, each point of $[1, n]$ is covered at least $m+1$ times by couples of intervals from B. Then

$$E(x, y^*) \geq \frac{m+1}{n}.$$

Suppose that Infiltrator employs the mixed strategy x^* and Guard plays the pure strategy y. Then it is clear that Guard will detect Infiltrator with probability at most $\frac{m+1}{n}$. So,

$$E(x^*, y) \leq \frac{m+1}{n}.$$

This completes the proof of Theorem 2.2.4.

Definition 2.2.2 *Let γ be a non-negative integer such that $n = \gamma m + \delta$, where $\delta \in [0, m-1]$.*

Remark 2.2.2 *(i) If $\alpha + \beta < m$ then*
(a) $\gamma = \alpha, \quad \delta = \alpha + \beta$ and hence $\gamma < \delta$,
(b) $(\delta+1)\gamma < (\gamma+1)\delta \leq n$,
(c) $\frac{(\gamma+1)\delta}{\delta+1} = \gamma + \frac{\delta-\gamma}{1+\delta}$.
(ii) If $\alpha+\beta \geq m$ then $\gamma = \alpha+1, \delta = \alpha+\beta-m$ and, hence, $\gamma > \delta$.

Proof (ia) By Definition 2.2.1 we have

$$n = (m+1)\alpha + \beta = m\alpha + \alpha + \beta.$$

Then, since $\alpha + \beta < m$, using Definition 2.2.1 implies (ia).
(ib) follows from (ia) and Definition 2.2.1. (ic) and (ii) are obvious. This completes the proof of Remark 2.2.2.

Theorem 2.2.5 *If $\alpha + \beta < m$ then*

$$v(m, 1, n) = \frac{\delta+1}{(\gamma+1)\delta}.$$

An optimal mixed strategy x^* of Infiltrator is to choose with equal probability a crossing point from the following set

$$A = \{i \in [1, n] : i = sm + t,\ s \in [0, \gamma],\ t \in [1, \delta]\}.$$

An optimal mixed strategy y^* of Guard is to choose with equal probability an interval from the following set of intervals $B = \{(M(p_i), K(\xi_i)), i \in [1, (\gamma + 1)\delta]\}$, where

$$p_i = \begin{cases} 1 + m\lfloor \frac{i}{\gamma+1} \rfloor & \textit{for } i \in [1, (\delta+1)\gamma - 1]\,, \\ 1 + (\gamma-1)m + \delta - \gamma & \textit{for } i = (\delta+1)\gamma \\ n - m + 1 & \textit{for } i \in [(\delta+1)\gamma + 1, (\gamma+1)\delta], \end{cases}$$

$$\xi_i = \begin{cases} 1 + \gamma m + \lfloor \frac{i}{\gamma} \rfloor & \textit{for } i \in [1, (\delta-\gamma)\gamma]\,, \\ 1 + \gamma m + \delta - \gamma & \\ +\lfloor \frac{i - (\delta-\gamma)\gamma}{\gamma+1} \rfloor & \textit{for } i \in [(\delta-\gamma)\gamma + 1, (\delta+1)\gamma]\,, \\ m(\gamma-1) + i - (\delta+1)\gamma & \textit{for } i \in [(\delta+1)\gamma + 1, (\gamma+1)\delta]\,. \end{cases} \tag{2.2}$$

Proof By Remark 2.2.2, (2.2) correctly defines the pure strategies of Guard. Assume that Guard adopts the mixed strategy y^* and Infiltrator plays the pure strategy x. It is easy to see that

- Each point $t \in [1 + m(\gamma-1), m(\gamma-1) + \delta - \gamma]$ is covered once by $\{K(\xi_i)\}$
- Each point $t \in [1 + m\gamma, m\gamma + \delta - \gamma]$ is covered γ times by $\{K(\xi_i)\}$
- Each point $t \in [1 + m\gamma + \delta - \gamma, n]$ is covered $\gamma + 1$ times by $\{K(\xi_i)\}$
- Each point $t \in [1, m(\gamma-1)]$ is covered $\delta + 1$ times by $\{M(p_i)\}$
- Each point $t \in [1+m(\gamma-1), m(\gamma-1)+\delta-\gamma]$ is covered δ times by $\{M(p_i)\}$
- Each point $t \in [1 + m(\gamma-1) + \delta - \gamma, m\gamma]$ is covered at least $\delta + 1$ times by $\{M(p_i)\}$
- Each point $t \in [1 + m\gamma, m\gamma + \delta - \gamma]$ is covered $\delta - \gamma + 1$ times by $\{M(p_i)\}$
- Each point $t \in [1 + m\gamma + \delta - \gamma, n]$ is covered $\delta - \gamma$ times by $\{M(p_i)\}$.

So, each point of $[1, n]$ is covered at least $\delta + 1$ times by couples of intervals from B. Then

$$E(x, y^*) \geq \frac{\delta+1}{(\gamma+1)\delta}.$$

Suppose that Infiltrator employs the mixed strategy x^* and Guard plays the pure strategy y. Note that the set A has the following structure. Let $I_i = [im + 1, (i+1)m], I^i = [1 + \delta + im, \delta + (i+1)m], i \in [0, \gamma - 1]$. Then $A = [1, n] \setminus \cup_{i \in [0,\gamma-1]} (I_i \cap I^i) = \cup_{i \in [0,\gamma-1]} (I_i \setminus I^i) \cup (I^{\gamma-1} \setminus I_{\gamma-1})$. Since $|I_i \cap I^i| = m - \delta$ we have $|A| = (\gamma+1)\delta$. It is clear that any integer interval consisting of m points covers at most δ points of A. It follows that Guard will detect Infiltrator with probability at most $\dfrac{\delta+1}{(\gamma+1)\delta}$. So,

$$E(x^*, y) \leq \frac{\delta+1}{(\gamma+1)\delta}.$$

This completes the proof of Theorem 2.2.5.

Theorem 2.2.6 *Let $m \in [2, n/2)$, then*

$$v(m,1,n) = \min\left\{\frac{m+1}{n}, \frac{\delta+1}{(\gamma+1)\delta}\right\}.$$

Proof It is easy to see that

$$\frac{m+1}{n} \ \{<,=,>\}\ \frac{\delta+1}{(\gamma+1)\delta} \quad \text{if} \quad \delta\ \{<,=,>\}\ \gamma.$$

By Assumption 2.2.1 the case $\delta = \gamma$ does not hold and by Remark 2.2.2 (ia) and (ii)

$$\delta\{<,>\}\gamma \quad \text{if} \quad \alpha+\beta\{\geq,<\}m.$$

So, Theorem 2.2.6 follows from Theorems 2.2.4 and 2.2.5.

We briefly discuss the case $k = 2$ below assuming that m is odd and α is a non-negative integer such that $n = \alpha(m+2) + 2\beta$, where $\beta \in [1,\ m+1]$.

Theorem 2.2.7 *Let $\alpha + \beta \geq m$ then*

$$v(m,2,n) = \frac{m+2}{n}.$$

An optimal mixed strategy x^* of Infiltrator is to choose with equal probability a crossing point from $[1,n]$. An optimal mixed strategy y^* of Guard is to choose with equal probability an interval from the following set of intervals $B = \{(M(p_i), K(\xi_i)), i \in [1,n]\}$, where
(a) if $n = 2m+2$, i.e. $\alpha = 0$, $\beta = m+1$, then

$$p_i = \begin{cases} 1 & \text{for } i \in [1, m+1], \\ n-m+1 & \text{for } i \in [m+2, n], \end{cases}$$

$$\xi_i = \begin{cases} 1+2(i-1) & \text{for } i \in [1, \frac{m+1}{2}], \\ m+1 & \text{for } i \in [\frac{m+3}{2}, \frac{3(m+1)}{2}], \\ m+2+2\{i-\frac{3(m+1)}{2}-1\} & \text{for } i \in [\frac{3(m+1)}{2}+1, n]. \end{cases}$$

In the remaining cases we have

$$p_i = \begin{cases} 1+m\lfloor\frac{i}{m+2}\rfloor & \text{for } i \in [1, n-\beta], \\ 1+m(\alpha+1) & \text{for } i \in [n-\beta+1, n] \end{cases}$$

and
(b) if $\alpha+\beta \geq m+1$ and $\alpha > 0$, then

$$\xi_i = \begin{cases} 1+\alpha m+2\lfloor\frac{i}{\tau}\rfloor & \text{for } i \in [1, m\tau], \\ 1+(\alpha+2)m+2\lfloor\frac{i-m\tau}{m+2}\rfloor & \text{for } i \in [1+m\tau, n], \end{cases}$$

where $\tau = m + 2 - \beta$,
(c) if $\alpha + \beta = m$ *and* $\alpha > 0$, *then*

$$\xi_i = \begin{cases} 1 + \alpha m \\ +2\left\lfloor \dfrac{i + n - \alpha(m+2) - \beta}{\tau} \right\rfloor & \text{for } i \in [1, \rho - \beta], \\ 2 + (\alpha+1)m \\ +2\left\lfloor \dfrac{i - \rho + 2\beta}{\tau} \right\rfloor & \text{for } i \in [\rho - \beta + 1, 2\rho - 2\beta], \\ (\alpha+1)m & \text{for } i \in [1 + 2\rho - 2\beta, \alpha(m+2)], \\ 2 + (\alpha+1)m \\ +2\left\lfloor \dfrac{i - \alpha(m+2)}{\tau} \right\rfloor & \text{for } i \in [1 + \alpha(m+2), \beta + \alpha(m+2)], \\ 1 + \alpha m \\ +2\left\lfloor \dfrac{i - \beta - \alpha(m+2)}{\tau} \right\rfloor & \text{for } i \in [1 + \beta + \alpha(m+2), n], \end{cases}$$

where $\rho = \tau(m-1)/2$.

Theorem 2.2.8 *If* $\alpha + \beta < m$ *and* $\theta \geq 0$, *where* $\theta = \lfloor \delta/2 \rfloor - \gamma + 1$ *then*

$$v(m, 2, n) = \frac{2 + \lfloor \delta/2 \rfloor}{(\gamma+1)(1 + \lfloor \delta/2 \rfloor)}.$$

An optimal mixed strategy x^* *of Infiltrator is to choose with equal probability a crossing point from the following set*

$$A = \left\{ i \in [1, n] : \begin{matrix} \text{there are } j \in [0, \gamma] \text{ and } s \in [0, \lfloor \delta/2 \rfloor] \text{ such that} \\ i = 1 + jm + 2s \end{matrix} \right\}.$$

An optimal mixed strategy y^* *of Guard is to choose with equal probability an interval from the following set of intervals* $B = \{(M(p_i), K(\xi_i)), i \in [1, (\gamma + 1)(1 + \lfloor \delta/2 \rfloor)]\}$, *where*

$$p_i = \begin{cases} 1 + m\left\lfloor \dfrac{i}{2 + \lfloor \delta/2 \rfloor} \right\rfloor & \text{for } i \in [1, \gamma(2 + \lfloor \frac{\delta}{2} \rfloor) - 1], \\ 1 + (\gamma - 1)m + 2\theta & \text{for } i = \gamma(2 + \lfloor \frac{\delta}{2} \rfloor), \\ n - m + 1 & \text{for } i \in [\gamma(2 + \lfloor \frac{\delta}{2} \rfloor) + 1, (\gamma+1)(1 + \lfloor \frac{\delta}{2} \rfloor)], \end{cases}$$

$$\xi_i = \begin{cases} 1 + \gamma m + 2\lfloor \frac{i}{\gamma} \rfloor & \text{for } i \in [1, \gamma\theta], \\ 1 + \gamma m + 2\theta \\ +2\left\lfloor \dfrac{i - \gamma\theta}{\gamma + 1} \right\rfloor & \text{for } i \in [\gamma\theta + 1, \gamma(1 + \lfloor \frac{\delta}{2} \rfloor) - 1], \\ n - 1 & \text{for } i \in [\gamma(1 + \lfloor \delta/2 \rfloor), \gamma(2 + \lfloor \frac{\delta}{2} \rfloor)], \\ 1 + (\gamma - 1)m \\ +2\{i - \gamma(2 + \lfloor \frac{\delta}{2} \rfloor) - 1\} & \text{for } i \in [\gamma(2 + \lfloor \frac{\delta}{2} \rfloor) + 1, (\gamma+1)(1 + \lfloor \frac{\delta}{2} \rfloor)]. \end{cases}$$

Let

$$U = \lfloor \tfrac{\delta}{2} \rfloor + \lfloor \tfrac{m-\delta-2}{2} \rfloor + 3,$$
$$D = (\gamma+1)(\lfloor \tfrac{\delta}{2} \rfloor + 1) + \gamma(\lfloor \tfrac{m-\delta-2}{2} \rfloor + 1).$$

Theorem 2.2.9 *If $\alpha+\beta<m$ and $\theta<0$ then*

$$v(m,2,n) = \frac{U}{D}.$$

An optimal mixed strategy x^* of Infiltrator is to choose with equal probability a crossing point from the following set

$$A = \left\{ i \in [1,n] : \begin{array}{c} \text{either there are } j \in [0,\gamma],\ s \in [0, \lfloor \delta/2 \rfloor] \text{ such that} \\ i = 1 + jm + 2s \text{ or there are } j \in [0,\gamma-1],\ s \in \\ [0, \lfloor (m-\delta-2)/2 \rfloor] \text{ such that } i = 2+\delta+jm+2s \end{array} \right\}.$$

An optimal mixed strategy y^* of Guard is to choose with equal probability an interval from the following set of intervals $B = \{(M(p_i), K(\xi_i)), i \in [1,D]\}$, where

$$p_i = 1 + m\lfloor \tfrac{i}{U} \rfloor, \quad i \in [1,D],$$
$$\xi_i = \begin{cases} 1 + (\gamma-1)m + 2\lfloor -\frac{i}{\theta} \rfloor & \text{for } i \in [1, -\frac{\theta(m-1)}{2}], \\ \gamma m + 2\left\lfloor \frac{i+\theta(m-1)/2}{U} \right\rfloor & \text{for } i \in [1 - \frac{\theta(m-1)}{2}, D]. \end{cases}$$

From Theorems 2.2.3- 2.2.9 we immediately obtain

Theorem 2.2.10 *Let $m \in (2, n/2)$ then*
(a) if m is even then $v(m,2,n) = v(m/2, 1, \lfloor n/2 \rfloor + 1)$,
(b) if m is odd then

$$v(m,2,n) = \begin{cases} \frac{m+2}{n}, & \text{for } \alpha+\beta \geq m, \\ \frac{\lfloor \delta/2 \rfloor + 2}{(\gamma+1)(\lfloor \delta/2 \rfloor + 1)} & \text{for } \alpha+\beta < m \text{ and } \theta \geq 0, \\ \frac{U}{D} & \text{for } \alpha+\beta < m \text{ and } \theta < 0. \end{cases}$$

2.3 Further Reading

The problems considered in Section 2.1 can be reformulated in terms of search problems. $\Gamma(m,n)$ can be described as a game between Searcher and Hider where Hider hides himself (or an object) in one point of the integer interval $[1,n]$. Searcher possesses search facilities enabling him to perform an attempt of search in any m consecutive points of the interval $[1,n]$. If Searcher looks through the point where Hider is then he is detected with certainty. Payoff to Searcher is 1 if Hider is found and 0 otherwise. This game was solved by Ruckle [86]. Also, he formulated a generalization of this game where Searcher has greater facilities allowing to search in m and k consecutive points simultaneously. Baston and Bostock [10] found solution of the continuous variant of the Ruckle's game for $m \geq n/2$. Garnaev [50] studied discrete version of the game for $k = 1, 2$. Ruckle [86] also formulated a generalization of the game for the case where Hider hides a big object occupying p consecutive points of $[1, n.]$ Denote this generalization of the game $\Gamma(m, n)$ and $\Gamma(m, k, n)$ by $\Gamma^{\mathrm{p}}(m, n)$ and $\Gamma^{\mathrm{p}}(m, k, n)$, respectively. It is easy to see that

$$\mathrm{val}(\Gamma^{\mathrm{p}}(m, n)) = \frac{1}{\lfloor n/(m + p - 1) \rfloor + 1}.$$

Solution of the game $\Gamma^{\mathrm{p}}(m, k, n)$ seems very difficult.

2.4 Number Hides Game

Ruckle [87] formulated a modification of the game $\Gamma^{p}(m, n)$ called Number Hides Game and denoted by $\Gamma^{\mathrm{p}}_{\mathrm{m}}(n)$ where the payoff to Searcher is the number of points in intersection of segments chosen by players. Baston, Bostock and Ferguson [15] solved this game. Professor Ruckle said me that the Number Hides Game was also solved by Zoroa and Zoroa independently from Baston, Bostock and Ferguson. IJGT (to its further discredit) rejected the paper after a long delay so they had to publish it in a Spanish journal.

Baston, Bostock and Ferguson proved the following two theorems 2.4.1-2.4.2.

Theorem 2.4.1 *Let $m \geq p$ and $n = Mm+\xi$, where $M \geq 1$ and $\xi \in [0, m-1]$. Then*

$$\mathrm{val}(\Gamma^{\mathrm{p}}_{\mathrm{m}}(n)) = \begin{cases} \dfrac{p(M+1) - \xi}{M(M+1)} & \text{for } \xi \leq p, \\ \dfrac{p}{M+1} & \text{for } \xi > p. \end{cases}$$

- *If $\xi = 0$ then optimal strategy x^* of Searcher is to choose with equal probability one of the following M intervals $\{[1 + im, (i + 1)m], i \in [0, M - 1]\}$. Optimal strategy y^* of Hider is to choose with equal probability one of the following M intervals $\{[1 + im, im + p], i \in [0, M - 1]\}$.*
- *If $\xi > 0$ then optimal strategy x^* of Searcher is to choose with probability $\dfrac{M-i}{M(M+1)}$ the interval $[1 + im, (i + 1)m]$ and with probability $\dfrac{i+1}{M(M+1)}$ the interval $[1+im+\xi, (i+1)m+\xi]$, where $i \in [0, M-1]$. Optimal strategy y^* of Hider is to choose with probability $\dfrac{M-i}{M(M+1)}$ the interval $[1+im, im+p]$ and with probability $\dfrac{i+1}{M(M+1)}$ the interval $[1 + im + \xi - p, (i + 1)m + \xi]$, where $i \in [0, M - 1]$.*

Proof (a) Straightforward.
(b) Let Hider use a pure strategy y saying him to choose the interval $[y, y + p - 1]$ and Searcher employs the strategy x^* then

- The intervals $[n-\xi+1, n]$ and $[1+im, im+\xi]$ for $i \in [0, M-1]$ will be covered by x^* with probability $\frac{1}{M+1}$
- The intervals $[1+\xi+im, (i+1)m]$ for $i \in [0, M-1]$ will be covered by x^* with probability $\frac{1}{M}$

So,

$$E(x^*, y) \geq \begin{cases} \frac{\xi}{M+1} + \frac{p-\xi}{M} = \frac{p(M+1)-\xi}{M(M+1)}, & \text{for } \xi \leq p, \\ \frac{p}{M+1}, & \text{for } \xi > p. \end{cases}$$

Let Searcher uses a pure strategy x saying him to choose the interval $[x, x+m-1]$ and Hider employs the strategy y^* then

(i) If $p \leq \xi$ then

- The intervals $[n-p+1, n]$ and $[1+im, im+p]$ for $i \in [0, M-1]$ will be covered by y^* with probability at most $\frac{1}{M+1}$
- The rest points of $[1, n]$ will not be covered by y^*

So,

$$E(x, y^*) \leq \frac{p}{M+1}.$$

(ii) If $p > \xi$ and $2p < m+\xi$ then

- The intervals $[1, p], [n-p+1, n]$ and $[1+im, \xi+im]$ for $i \in [1, M-1]$ will be covered by y^* with probability $\frac{1}{M+1}$
- The intervals $[1+\xi-p+im, im]$ for $i \in [1, M-1]$ will be covered by y^* with probability $\frac{i}{M(M+1)}$
- The intervals $[1+\xi+im, p+im]$ for $i \in [1, M-1]$ will be covered by y^* with probability $\frac{M-i}{M(M+1)}$,
- The rest points of $[1, n]$ will not be covered by y^*

So,

$$E(x, y^*) \leq \frac{p}{M+1} + \frac{p-\xi}{M(M+1)} = \frac{(M+1)p-\xi}{M(M+1)}.$$

(c) If $p > \xi$ and $2p \geq m+\xi$ then

- The intervals $[1, \xi+m-p], [n-\xi-m+p+1, n]$ and $[1+im, \xi+im]$ for $i \in [1, M-1]$ will be covered by y^* with probability $\frac{1}{M+1}$
- The intervals $[1+\xi+im, m-p+\xi+im]$ for $i \in [1, M-1]$ will be covered by y^* with probability $\frac{M-i}{M(M+1)}$
- The intervals $[1+m-p+\xi+im, p+\xi+im]$ for $i \in [0, M-1]$ will be covered by y^* with probability $\frac{1}{M}$

- The intervals $[1+p+\xi+im,\ (i+1)m]$ for $i \in [0,\ M-2]$ will be covered by y^* with probability $\frac{i+1}{M(M+1)}$

So,

$$E(x,y^*) \le \frac{\xi+m-p}{M+1} + \frac{2p-\xi-m}{M} + \frac{m-p}{M(M+1)} = \frac{(M+1)p-\xi}{M(M+1)}.$$

This completes the proof of Theorem 2.4.1.

In the similar manner the following theorem can be proved.

Theorem 2.4.2 *Let $m < p$ then*
(a) If $m+n \le 2p$, then $\mathrm{val}(\Gamma^{\mathrm{p}}_{\mathrm{m}}(n)) = m$ and any strategy is optimal for Hider. Optimal strategy x^* of Searcher is to choose the interval

$$\left[[\frac{p-m}{2}]+1, [\frac{p-m}{2}]+m\right],$$

where $[s]$ is the greatest integer less or equal to s.
(b) If $m+n > 2p$, let $n = Np+\xi$, where $N \ge 1$ and $\xi \in [0,\ p-1]$. Then

$$\mathrm{val}(\Gamma^{\mathrm{p}}_{\mathrm{m}}(n)) = \begin{cases} \frac{mN+p-\xi}{N(N+1)} & \text{for } \xi \ge p-m, \\ \frac{m}{N} & \text{for } \xi < p-m. \end{cases}$$

- *If $\xi = 0$ then optimal strategy x^* of Searcher is to choose with equal probability one of the following N intervals $\{[1+ip,\ ip+m], i \in [0,\ N-1]\}$. Optimal strategy y^* of Hider is to choose with equal probability one of the following N intervals $\{[1+ip,\ (i+1)p], i \in [0,\ N-1]\}$*
- *If $\xi > 0$ then optimal strategy x^* of Searcher is to choose with probability $\frac{N-i}{N(N+1)}$ the interval $[1+ip,\ ip+m]$ and with probability $\frac{i+1}{N(N+1)}$ the interval $[1+ip+\xi-m,\ (i+1)p+\xi]$, where $i \in [0,\ N-1]$. Optimal strategy y^* of Hider is to choose with probability $\frac{N-i}{N(N+1)}$ the interval $[1+ip,\ (i+1)p]$ and with probability $\frac{i+1}{N(N+1)}$ the interval $[1+ip+\xi,\ (i+1)p+\xi]$, where $i \in [0,\ N-1]$*

2.5 Matrix Search Games

The games considered in the previous sections are matrix ones. For example, the game $\Gamma(m,n)$ has the following $(n-m) \times n$-payoff matrix

$$a_{ij} = \begin{cases} 1 & \text{for } 1 \le i \le j \le i+m-1, \\ 0 & \text{otherwise}. \end{cases}$$

The game $\Gamma^{p}(m,n)$ has the following $(n-m)\times(n-p)$-payoff matrix
(a) for $m > p$

$$a_{ij} = \begin{cases} p-i+j & \text{for } i-p+1 \le j \le i\,, \\ p & \text{for } i \le j \le m+i-p\,, \\ m+i-j & \text{for } m+i-p \le j \le m+i-1\,, \\ 0 & \text{otherwise}\,, \end{cases}$$

(b) for $m \le p$

$$a_{ij} = \begin{cases} m-j+i & \text{for } j-m+1 \le i \le j\,, \\ m & \text{for } j \le i \le p+j-m\,, \\ p+j-i & \text{for } p+j-m \le i \le p+j-1\,, \\ 0 & \text{otherwise}\,. \end{cases}$$

Among matrix games the simplest one seems the following game. Player 1 chooses a number, say i, from the integer interval $[1, n]$. Player 2 also chooses a number there, say j. The payoff to player 1 is $|i-j|$. So, this game is a matrix game with the following payoff matrix

$$a_{ij} = |i-j| \quad \text{for} \quad i,\ j \in [1,\ n]\,.$$

It is easy to see that the value of this game is $(n-1)/2$. The optimal strategies x^* and y^* of player 1 and 2, respectively, are ginen by

$$x_i^* = \begin{cases} 1/2 & \text{for } i=1 \text{ or } i=n\,, \\ 0 & \text{otherwise}\,, \end{cases}$$

if $n = 2k+1$ then

$$y_i^* = \begin{cases} 1 & \text{for } i=k+1\,, \\ 0 & \text{otherwise}\,, \end{cases}$$

if $n = 2k$ then

$$y_i^* = \begin{cases} 1/2 & \text{for } i=k \text{ or } i=k+1\,, \\ 0 & \text{otherwise}\,. \end{cases}$$

The simplest search matrix game seems the following one. There are n boxes. Hider hides an object in one of n boxes. Searcher chooses one of these boxes to perform one attempt of search there. If Searcher searches in box i where Hider is then the probability of his detection is α_i, where $\alpha_i \in (0,\ 1)$, and 0 otherwise. The payoff to Searcher is 1 if Hider is detected and 0 otherwise. Then, the game has the following $n \times n$ payoff matrix

$$a_{ij} = \begin{cases} \alpha_i & \text{for } i=j\,, \\ 0 & \text{otherwise}\,. \end{cases}$$

The value v of the game is given by

$$v = \frac{1}{\sum_{j=1}^{n} 1/\alpha_j}.$$

The optimal strategies $x^* = (x_1^*, \ldots, x_n^*)$ and $y^* = (y_1^*, \ldots, y_n^*)$ of Searcher and Hider, respectively, are given by

$$x_i^* = y_i^* = \frac{1/\alpha_i}{\sum_{j=1}^{n} 1/\alpha_j} \quad \text{for} \quad i \in [1, n].$$

Sakaguchi [91] suggested to consider a variant of the game where cheking of box i costs c_i and the reward to Searcher for finding Hider is R. The payoff to Searcher is the search's cost. So, the game has the following $n \times n$ payoff matrix

$$a_{ij} = \begin{cases} \alpha_i R - c_i & \text{for } i = j, \\ -c_i & \text{otherwise}. \end{cases}$$

Without loss of generality we can assume that

$$c_1 = \min_{i \in [1,n]} c_i.$$

Theorem 2.5.1 *The value of the game v, and the optimal strategy $x^* = (x_1^*, \ldots, x_n^*)$ and $y^* = (y_1^*, \ldots, y_n^*)$ of Searcher and Hider are given as follows*

(a) if

$$\sum_{j=2}^{n} \frac{c_j - c_1}{\alpha_j} \le R \tag{2.3}$$

then

$$v = \frac{R - \sum_{j=1}^{n} (c_j/\alpha_j)}{\sum_{j=1}^{n} (1/\alpha_j)},$$

$$x_i^* = \frac{1/\alpha_i}{\sum_{j=1}^{n} (1/\alpha_j)},$$

$$y_i^* = \frac{c_i + v}{R\alpha_i} \quad \textit{for} \quad i \in [1, n], \tag{2.4}$$

(b) if

$$\sum_{j=k+1}^{n} \frac{c_j - c_1}{\alpha_j} \le R < \sum_{j=k}^{n} \frac{c_j - c_1}{\alpha_j}, \tag{2.5}$$

where $k \in [2, n-1]$ *then*

$$\begin{aligned} v &= -c_1 \\ x_i^* &= \begin{cases} 1 & \textit{for } i = 1, \\ 0 & \textit{for otherwise}, \end{cases} \\ y_i^* &= \begin{cases} 0 & \textit{for } i \in [1,\, k-1], \\ 1 - \displaystyle\sum_{j=k+1}^{n} \frac{c_j - c_1}{R\alpha_j} & \textit{for } i = k, \\ \dfrac{c_i - c_1}{R\alpha_i} & \textit{for } i \in [k+1,\, n], \end{cases} \end{aligned} \tag{2.6}$$

(c) if

$$R \le \frac{c_n - c_1}{\alpha_n}, \tag{2.7}$$

then

$$\begin{aligned} v &= -c_1, \\ x^* &= (1, 0, \ldots, 0), \\ y^* &= (0, 0, \ldots, 1). \end{aligned} \tag{2.8}$$

Proof v is the value of the game, and x^* and y^* are optimal strategy of Searcher and Hider, respectively, if and only if the following inequalities hold

$$\begin{array}{lll} (\alpha_1 R - c_1)x_1^* & -c_2 x_2^* - \ldots -c_n x_n^* & \ge v, \\ \ldots & \ldots & \\ -c_1 x_1^* & -c_2 x_2^* - \ldots +(\alpha_n R - c_n)x_n^* & \ge v, \\ (\alpha_1 R - c_1)y_1^* & -c_1 y_2^* - \ldots -c_1 y_n^* & \le v, \\ \ldots & \ldots & \\ -c_n y_1^* & -c_n y_2^* - \ldots +(\alpha_n R - c_n)y_n^* & \le v. \end{array}$$

These inequalities are equivalent to the following ones

$$\begin{array}{ll} \alpha_i R x_i^* - \sum_{j=1}^{n} c_j x_j^* \ge v, & \\ \alpha_i R y_i^* \quad -c_i \quad \le v, & \text{for } i \in [1, n]. \end{array} \tag{2.9}$$

(a) Let v, x^* and y^* be given by (2.4). Then, by (2.3)

$$\frac{1 - \sum_{j=1}^{n} (c_j/(R\alpha_j))}{\sum_{j=1}^{n} (1/(R\alpha_j))} \ge -c_1.$$

So,

$$v \geq -c_1 .$$

This implies that y^* given by (2.4) is a probability vector. It is easy to see that the inequalities (2.9) hold as equalities. Hence, v is the value of the game, and x^* and y^* are optimal strategy of Searcher and Hider, respectively.

(b) Let v, x^* and y^* be given by (2.6). Then, by (2.5) y^* given by (2.6) is a probability vector. Using (2.6) implies that

$$\alpha_i R x_i^* - \sum_{j=1}^{n} c_j x_j^* = \left\{ \begin{array}{rl} \alpha_1 R - c_1 \text{ for } & i = 1, \\ -c_1 \text{ for } & i > 1 \end{array} \right\} \geq -c_1 = v . \qquad (2.10)$$

By (2.5) and (2.6),

$$\alpha_i R y_i^* - c_i = \left\{ \begin{array}{rl} -c_i \text{ for } & i \in [1, k-1] \\ \alpha_k R \left(1 - \sum_{j=k+1}^{n} \frac{c_j - c_1}{R\alpha_j} \right) - c_k \text{ for } & i = k \\ -c_1 \text{ for } & i \in [k+1, n] \end{array} \right\}$$
$$\leq -c_1 = v.$$

Now (2.9) follows. So, v is the value of the game, and x^* and y^* are optimal strategy of Searcher and Hider, respectively.

(c) Let v, x^* and y^* be given by (ref32f6). Then, by (2.7) that

$$\alpha_i R y_i^* - c_i = \left\{ \begin{array}{rl} -c_i \text{ for } & i \in [1, n-1], \\ \alpha_n R - c_n \text{ for } & i = n \end{array} \right\} \leq -c_1 = v .$$

Then (2.10) implies that the inequalities (2.9) hold. This completes the proof of Theorem 2.5.1.

2.6 Helicopter Versus Submarine Game

Consider the following Helicopter and Submarine game. The game is played on an integer unterval $[1, n]$. Submarine chooses one of n possible possition on $[1, n]$ and submerges at the chosen point. Helicopter, having a bomb, does not know the submarine's location. It can drop the bomb at any position. If the bomb explodes at point i and Submarine is at point $i-1$, or i, or $i+1$, then the probability of its destruction is α, or 1, or α, respectively, where $\alpha \in (0, 1)$. The payoff to Helicopter is 1 if Submarine is destructed and 0 otherwise. This game has the following $n \times n$ payoff matrix to Helicopter

$$\begin{pmatrix} 1 & \alpha & & & & \\ \alpha & 1 & \alpha & & & \\ & \alpha & 1 & \alpha & & \\ & & \cdots & \cdots & \cdots & \\ & & & \alpha & 1 & \alpha \\ & & & & \alpha & 1 \end{pmatrix}.$$

Theorem 2.6.1 *(a) Let $\alpha < 1/2$ then the value v of the game is given by*

$$\frac{1}{v} = \frac{1}{1+2\alpha}\left(n + \frac{2A(A^n - 1)}{(1-A)(1+A^{n+1})}\right), \tag{2.11}$$

where

$$A = -\frac{1+\sqrt{1-4\alpha^2}}{2\alpha}.$$

The optimal strategies $x^* = (x_1^*, \ldots, x_n^*)$ and $y^* = (y_1^*, \ldots, y_n^*)$ of Helicopter and Submarine, respectively, are given as follows:

$$x_i^* = y_i^* = v z_i, \quad \textit{for} \quad i \in [1, n], \tag{2.12}$$

where

$$z_i = \frac{1}{1+2\alpha}\left(1 + \frac{1-B^{n+1}}{B^{n+1}-A^{n+1}}A^i + \frac{A^{n+1}-1}{B^{n+1}-A^{n+1}}B^i\right) \quad \textit{for} \quad i \in [0, n+1]$$

and

$$B = \frac{-1+\sqrt{1-4\alpha^2}}{2\alpha}.$$

(b) Let $\alpha = 1/2$ then the value v of the game is given by

$$v = \begin{cases} \dfrac{1}{k+1} & \textit{for } n = 2k+1, \\ \dfrac{2k+1}{2k(k+1)} & \textit{for } n = 2k. \end{cases}$$

The optimal strategies $x^* = (x_1^*, \ldots, x_n^*)$ and $y^* = (y_1^*, \ldots, y_n^*)$ of Helicopter and Submarine, respectively, are the following:
(i) if $n = 2k+1$ then for $i \in [1, n]$

$$x_i^* = y_i^* = \begin{cases} \dfrac{1}{k+1} & \textit{for } i = 2j+1 \textit{ and } j \in [0, k-1], \\ 0 & \textit{otherwise}, \end{cases}$$

(ii) if $n = 2k$ then for $i \in [1, n]$ and

$$x_i^* = y_i^* = \begin{cases} \dfrac{2j}{2k(k+1)} & \textit{for } i = 2j, \\ \dfrac{n-2j+2}{2k(k+1)} & \textit{for } i = 2j-1. \end{cases}$$

Proof First note that v is the value of the game, and x^* and y^* are optimal strategies of Searcher and Hider, respectively, if and only if the following inequalities hold

$$
\begin{aligned}
&x_1^* + \alpha x_2^* && \geq v\,,\\
&\alpha x_1^* + x_2^* \quad + \alpha x_3^* && \geq v\,,\\
&\qquad\quad \dots \qquad \dots && \\
&\qquad\quad \alpha x_{n-1}^* + \alpha x_n^* && \geq v\,,\\
&y_1^* + \alpha y_2^* && \leq v\,,\\
&\alpha y_1^* + y_2^* \quad + \alpha y_3^* && \leq v\,,\\
&\qquad\quad \dots \qquad \dots && \\
&\qquad\quad \alpha y_{n-1}^* + \alpha y_n^* && \leq v\,.
\end{aligned}
\tag{2.13}
$$

(a) It is clear that

$$AB = 1\,,$$

and A and B are the roots of the equation

$$\alpha\xi^2 + \xi + \alpha = 0\,.$$

Then simple computation shows that

$$\alpha z_{i-1} + z_i + \alpha z_{i+1} = 1 \text{ for } i \in [1, n]\,, \tag{2.14}$$

$$z_0 = z_{n+1} = 0 \tag{2.15}$$

and

$$\sum_{i=1}^{n} z_i = v\,,$$

where v is given by (2.11). From (2.14) and (2.15) we have that

$$
\begin{aligned}
&z_n = a_n\,,\\
&z_i = a_i - b_i z_{i+1} \text{ for } i = n-1, n-2, \dots, 1\,,
\end{aligned}
\tag{2.16}
$$

where

$$
\begin{aligned}
&b_1 = \alpha, && a_1 = 1,\\
&b_i = \frac{\alpha}{1 - \alpha b_{i-1}}\,, && a_i = \frac{1 - \alpha a_{i-1}}{1 - \alpha b_{i-1}}, \quad \text{for} \quad i \in [2, n]\,.
\end{aligned}
$$

Now we shall prove by induction on i that

$$0 < b_i < a_i \leq 1 \text{ for } i \in [1, n]\,. \tag{2.17}$$

For $i = 1$, (2.17) follows from the initial condition. Assume that it holds for $i = j$. Prove that then it also holds for $i = j + 1$. The assumption that $b_j < 1$ implies that

$$b_{j+1} = \frac{\alpha}{1 - \alpha b_j} > 0\,.$$

Since $\alpha < 1/2$, by the assumption that $a_j \leq 1$ we have that

$$\alpha < 1 - \alpha a_j\,.$$

So,

$$b_{j+1} = \frac{\alpha}{1-\alpha b_j} < \frac{1-\alpha a_j}{1-\alpha b_j} = a_{j+1}\,.$$

The assumption that $b_j < a_j$ implies that

$$a_{j+1} = \frac{1-\alpha a_j}{1-\alpha b_j} < \frac{1-\alpha b_j}{1-\alpha b_j} = 1\,.$$

Thus, (2.17) follows. By (2.16) and (2.17),

$$z_i \geq 0 \quad \text{for} \quad i \in [1, n]\,.$$

So, for v, and the strategies x^* and y^* given by (2.11) and (2.12), respectively, the inequalities (2.13) hold as equalities. This completes the proof of (a).

(b) can be checked by straitforward computation.

Baston and Bostock [13] studied a continuous variant of Helicopter and Submarine game where Submarine would be destructed with certainty within destractive radius from point of the bomb's explosion. Garnaev [46] investigated the Helicopter and Submarine game when the probability of destruction depended on the distance between Submarine and the point of the bomb's explosion.

Problems

Solve the game $\Gamma^p(m, 1, n)$ of search of a segment consisting of p points by a point and a segment consisting of m points for $m < n/2$. Of course, for $m \geq n/2$ quite analogously to Theorem 2.2.1 we can find that

$$\text{val}(\Gamma^p(m, 1, n)) = \frac{\lfloor (n-m)/p \rfloor + 2}{2(\lfloor (n-m)/p \rfloor + 1)}\,.$$

Solve the Helicopter and Submarine game for $\alpha < 1/2$. For example, for $n \in [3, 5]$ we have that

	v	x^*	y^*
3	α	$(0,1,0)$	$(\alpha, 0, \bar{\alpha})$
4	$\dfrac{1+\alpha-\alpha^2}{2(2-\alpha)}$	$\dfrac{(1, \bar{\alpha}, \bar{\alpha}, 1)}{2(2-\alpha)}$	$\dfrac{(1, \bar{\alpha}, \bar{\alpha}, 1)}{2(2-\alpha)}$
5	$\dfrac{\alpha}{1+\alpha}$	$\dfrac{(\alpha, 1, 0, 1, \alpha)}{2(1+\alpha)}$	$\dfrac{(\alpha, 0, \bar{\alpha}, 0, \alpha)}{1+\alpha}$

where $\bar{\alpha} = 1 - \alpha$. For $n = 6$ we have that

$$v = \begin{cases} \frac{\alpha}{2}, & \text{for } \alpha \geq \frac{-1+\sqrt{5}}{2}, \\ \frac{1+\alpha-2\alpha^2-\alpha^3}{2(3-2\alpha-2\alpha^2)}, & \text{otherwise,} \end{cases}$$

$$x^* = \frac{(0,\,1,\,0,\,0,\,1,\,0)}{2},\, y^* = \frac{(\alpha,\,0,\,\bar{\alpha},\,\bar{\alpha},\,0,\,\alpha)}{2} \quad \text{for} \quad \alpha \geq \frac{-1+\sqrt{5}}{2},$$
$$x^* = y^* = \frac{(1-\alpha^2,\,\bar{\alpha}-\alpha^2,\,\bar{\alpha},\,\bar{\alpha},\bar{\alpha}-\alpha^2,\,1-\alpha^2)}{2(3-2\alpha-2\alpha^2)} \quad \text{otherwise}.$$

For $n = 7$ the value of the game is given by

$$v = \begin{cases} \frac{\alpha}{2-\alpha+2\alpha^2} & \text{for } \alpha < \frac{-1+\sqrt{5}}{2}, \\ \frac{1-4\alpha^2+2\alpha^4}{7-12\alpha-6\alpha^2+8\alpha^3} & \text{otherwise}. \end{cases}$$

The optimal strategies of Helicopter and Submarine are given by for $\alpha < \frac{-1+\sqrt{5}}{2}$

$$x^* = v \times (\alpha,\, \bar{\alpha}/\alpha,\, 0,\, 1,\, 0,\, \bar{\alpha}/\alpha,\, \alpha),$$
$$y^* = v \times (1,\, 0,\, \bar{\alpha}/\alpha,\, 2\alpha-1,\, \bar{\alpha}/\alpha,\, 0,\, 1),$$

for $\alpha \geq \frac{-1+\sqrt{5}}{2}$

$$x_1^* = x_7^* = y_1^* = y_7^* = \frac{1-\alpha-2\alpha^2+\alpha^3}{7-12\alpha-6\alpha^2+8\alpha^3},$$
$$x_2^* = x_6^* = y_2^* = y_6^* = \frac{(1-\alpha)^2(1-2\alpha)}{7-12\alpha-6\alpha^2+8\alpha^3},$$
$$x_3^* = x_5^* = y_3^* = y_5^* = \frac{(1-\alpha)(1-\alpha-\alpha^2)}{7-12\alpha-6\alpha^2+8\alpha^3},$$
$$x_4^* = y_4^* = \frac{1-2\alpha}{7-12\alpha-6\alpha^2+8\alpha^3}.$$

It is interesting to note that for $n = 6$ the inequalities (2.13) hold as equalities for $\alpha < \frac{-1+\sqrt{5}}{2}$ but for $n = 7$ they hold as equalities for $\alpha \geq \frac{-1+\sqrt{5}}{2}$.

3 Allocation Games

3.1 One-Sided Allocation Game Without Search Cost

Consider the following zero-sum one-sided allocation game on integer interval $[1, n]$. Hider selects one of the n points and hides there. Searcher seeks Hider by dividing the given total continuous search effort X and allocating it in each point. Each point i is characterized by two detection parameters $\lambda_i > 0$ and $\alpha_i \in (0,1)$ such that $\alpha_i(1-\exp(-\lambda_i z))$ is the probability that a search of point i by Searcher with an amount of search effort z will discover Hider if he is there. The payoff to Searcher is 1 if Hider is detected and 0 otherwise. A strategy of Searcher and Hider can be represented by $x = (x_1, \ldots, x_n)$ and $y = (y_1, \ldots, y_n)$, respectively, where y_i is the probability that Hider hides in box i and x_i is the amount of effort allocated in box i by Searcher, where $x_i \geq 0$ for $i \in [1,n]$ and $\sum_{i=1}^{n} x_i = X$. So, the payoff to Searcher if Searcher and Hider employ strategies x and y, respectively, is given by

$$M(x,y) = \sum_{i=1}^{n} \alpha_i y_i (1 - \exp(-\lambda_i x_i)) .$$

Without loss of generality we can consider that

$$\alpha_1 = \min_{i \in [1,n]} \alpha_i .$$

Remark 3.1.1 *Since M is concave in x for fixed y and linear in y for fixed x, the Kuhn - Tucker Theorem (Mangasarian [71]) implies that x^* and y^* are optimal strategies of the players if and only if there are non - negative ω and ν such that*

$$M_i(x^*, y^*) = \frac{\partial}{\partial x_i} M(x^*, y^*) = \alpha_i \lambda_i y_i^* \exp(-\lambda_i x_i^*) \begin{cases} = \omega & \text{for } x_i^* > 0, \\ \leq \omega & \text{for } x_i^* = 0 \end{cases}$$

and

$$\alpha_i (1 - \exp(-\lambda_i x_i^*)) \begin{cases} = \nu & \text{for } y_i^* > 0, \\ \leq \nu & \text{for } y_i^* = 0 . \end{cases}$$

Theorem 3.1.1 *Let ν_* be the unique root in $(0, \alpha_1)$ of the following equation*

$$\varphi(\nu) = X .$$

where

$$\varphi(\nu) = \sum_{j=1}^{n} \frac{1}{\lambda_j} \ln\left(\frac{\alpha_j}{\alpha_j - \nu}\right).$$

Let $x^* = (x_1^*, \dots, x_n^*), y^* = (y_1^*, \dots, y_n^*)$ be such that

$$\begin{aligned} x_i^* &= \frac{1}{\lambda_i} \ln\left(\frac{\alpha_i}{\alpha_i - \nu_*}\right), \\ y_i^* &= \frac{1/(\lambda_i(\alpha_i - \nu_*)}{\sum_{j=1}^{n}(1/(\lambda_j(\alpha_j - \nu_*))} \quad \text{for } i \in [1, n]. \end{aligned} \tag{3.1}$$

Then x^* and y^* are the optimal strategies of Searcher and Hider, respectively, and ν_* is the value of the game.

Proof Let x^* and y^* be given by (3.1). By definition of ν_*, (3.1) correctly defines strategies of players. It is clear that

$$\alpha_i(1 - \exp(-\lambda_i x_i^*)) = \nu_* \quad \text{for} \quad i \in [1, n].$$

So,

$$M(x^*, y) = \nu_* \sum_{i=1}^{n} y_i = \nu_* \quad \text{for any } y.$$

Also, for $i \in [1, n]$

$$\begin{aligned} M_i(x^*, y^*) &= \lambda_i y_i^* \alpha_i \exp(-\lambda_i x_i^*) \\ &= \lambda_i y_i^* (\alpha_i - \nu_*) = \frac{1}{\sum_{j=1}^{n} 1/(\lambda_j(\alpha_j - \nu_*))} = \omega_* > 0. \end{aligned}$$

Now the result follows from Remark 3.1.1.

3.2 One-Sided Allocation Game with Search Cost

Consider the variant of the allocation game where Searcher has to pay for spent search efforts. Let R be a reward of Searcher for finding Hider and $C\sum_{i=1}^{n} x_i$, where $C > 0$, is a cost of spending $\sum_{i=1}^{n} x_i$ search efforts by Searcher. We assume that $\sum_{i=1}^{n} x_i \le X$. The payoff to Searcher is reward minus search's cost. So, the payoff to Searcher if Searcher and Hider employ strategies x and y, respectively, is given by

$$M(x, y) = R \sum_{i=1}^{n} \alpha_i y_i (1 - \exp(-\lambda_i x_i)) - C \sum_{i=1}^{n} x_i.$$

In this game Searcher trying to maximize his reward faces the alternatives of using all resources, part of the resources or not searching at all.

Remark 3.2.1 *Since M is concave in x for fixed y and linear in y for fixed x, the Kuhn - Tucker Theorem (Mangasarian [71]) implies that x^* and y^* are optimal strategies of the players if and only if there are non - negative ω and ν such that*

$$M_i(x^*, y^*) = \frac{\partial}{\partial x_i} M(x^*, y^*) = \alpha_i \lambda_i y_i^* \exp(-\lambda_i x_i^*) - C \begin{cases} = \omega & \text{for } x_i^* > 0\,, \\ \leq \omega & \text{for } x_i^* = 0 \end{cases}$$

and

$$R\alpha_i(1 - \exp(-\lambda_i x_i^*)) \begin{cases} = \nu & \text{for } y_i^* > 0\,, \\ \leq \nu & \text{for } y_i^* = 0\,, \end{cases}$$

where

$$\omega \begin{cases} \geq 0 & \text{for } \sum_{i=1}^n x_i^* = X\,, \\ = 0 & \text{otherwise}\,. \end{cases}$$

Theorem 3.2.1 *(a) If there is $i \in [1, n]$ such that $R\alpha_i\lambda_i \leq C$ then let*

$$\begin{aligned} v &= 0\,, \\ x_j^* &= 0 \text{ for } j \in [1, n]\,, \\ y_j^* &= \begin{cases} 1 & \text{for } j = i\,, \\ 0 & \text{otherwise}\,. \end{cases} \end{aligned} \tag{3.2}$$

(b) If

$$R\alpha_i\lambda_i > C \quad \text{for} \quad i \in [1, n] \tag{3.3}$$

and

$$\sum_{j=1}^n \frac{C}{R\lambda_j \alpha_j} \geq 1$$

then let

$$\begin{aligned} &v = 0\,, \\ &x_j^* = 0 \text{ for } j \in [1, n]\,, \\ &y_j^* \geq 0 \text{ for } j \in [1, n] \text{ such that } R\lambda_j\alpha_j y_j^* \leq C \text{ and } \sum_{j=1}^n y_j^* = 1\,. \end{aligned} \tag{3.4}$$

(c) If (3.3) holds and

$$\sum_{j=1}^n \frac{C}{R\lambda_j \alpha_j} < 1 \tag{3.5}$$

and

$$\psi(\nu_*) < 1, \tag{3.6}$$

where ν_* is given in Theorem 3.1.1 and

$$\psi(\nu) = \sum_{j=1}^{n} \frac{C}{R\lambda_j(\alpha_j - \nu)},$$

then let x^*, y^* be given by (3.1) and

$$v = R\nu_* - C\varphi(\nu_*). \tag{3.7}$$

(d) If (3.3) and (3.5) hold and (3.6) does not hold then let

$$\begin{aligned} v &= R\nu^* - C\varphi(\nu^*), \\ x_i^* &= \frac{1}{\lambda_i} \ln\left(\frac{\alpha_i}{\alpha_i - \nu^*}\right), \\ y_i^* &= \frac{1/(\lambda_i(\alpha_i - \nu^*)}{\sum_{j=1}^{n}(1/(\lambda_j(\alpha_j - \nu^*))} \quad \text{for} \quad i \in [1, n], \end{aligned} \tag{3.8}$$

where ν^* is the unique root in $(0, \alpha_1)$ of the following equation

$$\psi(\nu) = 1.$$

Then x^* and y^* are the optimal strategies of Searcher and Hider, respectively, and v is the value of the game.

Proof (a) Let v, x^* and y^* be given by (3.2). Then, for $j \in [1, n]$

$$M_j(x^*, y^*) = \begin{Bmatrix} R\alpha_i\lambda_i - C, \text{for } j = i \\ -C, \text{otherwise} \end{Bmatrix} \le 0.$$

Also,

$$R\alpha_i(1 - \exp(-\lambda_i x_i^*)) = 0 \quad \text{for} \quad i \in [1, n]. \tag{3.9}$$

So,

$$M(x^*, y) = 0 \quad \text{for any strategy} \quad y.$$

Then, by Remark 3.2.1 with $\omega = 0 = \nu$, x^* and y^* given by (3.2) are the optimal strategies of Searcher and Hider, respectively, and 0 is the value of the game.

(b) Let v, x^* and y^* be given by (3.4). Then, for $j \in [1, n]$

$$M(x^*, y^*) = R\alpha_j\lambda_j y_j^* - C \le 0.$$

Also, it is clear that (3.9) holds. Then, by Remark 3.2.1 with $\omega = 0 = \nu$, x^* and y^* given by (3.4) are the optimal strategies of Searcher and Hider, respectively, and 0 is the value of the game.

(c) Let v, x^* and y^* be given by (3.7) and (3.1), respectively. Then

$$R\alpha_i(1 - \exp(-\lambda_i x_i^*)) = R\nu_* \quad \text{for} \quad i \in [1, n].$$

So,

$$M(x^*, y) = R\nu_* - C\varphi(\nu_*) \quad \text{for any strategy} \quad y.$$

Also, for $i \in [1, n]$

$$\begin{aligned} M_i(x^*, y^*) &= R\lambda_i y_i^* \alpha_i \exp(-\lambda_i x_i^*) - C \\ &= R\lambda_i y_i^* (\alpha_i - \nu_*) - C = C\left(\frac{1}{\psi(\nu_*)} - 1\right) > 0. \end{aligned}$$

Thus, since $\sum_{i=1}^n x_i^* = X$, Remark 3.2.1 with $\nu = R\nu_*$ and $\omega = C(1/\psi(\nu_*) - 1)$, implies x^* and y^* given by (3.1) are the optimal strategies of Searcher and Hider, respectively, and v given by (3.7) is the value of the game.

(d) Let v, x^* and y^* be given by (3.8). Then,

$$R\alpha_i(1 - \exp(-\lambda_i x_i^*)) = R\nu^* \quad \text{for} \quad i \in [1, n].$$

So,

$$M(x^*, y) = R\nu^* - C\varphi(\nu^*) \quad \text{for any strategy} \quad y.$$

Also, for $i \in [1, n]$

$$M_i(x^*, y^*) = R\lambda_i y_i^* \alpha_i \exp(-\lambda_i x_i^*) - C = 0.$$

Thus, since $\sum_{i=1}^n x_i^* \le X$, Remark 3.2.1 with $\nu = R\nu^*$ and $\omega = 0$, implies that x^* and y^* given by (3.8) are the optimal strategies of Searcher and Hider, respectively, and v is the value of the game. This completes the proof of Theoren 3.2.1.

3.3 Further Reading

Danskin [26] studied the one-sided zero-sum allocated game. He formulated it as an antiballistic missile defence problem. Nakai [76] found closed form solution of the game where $\alpha_i = 1, i \in [1, n]$. Iida, Hohzaki and Satō [55] investigated the game where Searcher has to pay for his search efforts and they found explicit solution of this game.

In this Section we consider a generalization of the one-sided allocation game for the case where for detecting of Hider in box i Searcher gets a reward R_i and for not finding him he gets fined r_i. So, the payoff to Searcher if Searcher and Hider employ strategies x and y, respectively, is given by

$$M(x,y) = \sum_{i=1}^{n} y_i \left(R_i(1-\exp(-\lambda_i x_i)) - r_i \exp(-\lambda_i x_i)\right) .$$

where $\sum_{i=1}^{n} x_i = X, \sum_{i=1}^{n} y_i = Y$ and $x_i \geq 0, y_i \geq 0$ for $i \in [1,n]$. Without loss of generality we can consider that

$$r_1 \leq r_2 \leq \ldots \leq r_n .$$

Let $1 = s(1) < s(2) < \ldots < s(m+1) = n+1$ be such that

$$\begin{aligned} r_{s(1)} = \ldots = r_{s(2)-1} < r_{s(2)} = \ldots = r_{s(3)-1} < \ldots \\ < r_{s(m)} = \ldots = r_{s(m+1)-1} . \end{aligned} \tag{3.10}$$

Let

$$R_* = \min\{R_1, \ldots, R_n\}$$

and

$$\varphi_k(\omega) = \sum_{j=s(k)}^{n} \frac{1}{\lambda_j} \ln\left(\frac{R_j + r_j}{R_j - \omega}\right) .$$

Remark 3.3.1 *(a)* $\varphi_1(R_* - 0) = \infty$,
(b) $\varphi_k(-r_{s(k)}) = \varphi_{k+1}(-r_{s(k)})$ *for* $k \in [1, m-1]$,
(c) $\varphi_m(-r_{s(m)}) = 0$,
(d) φ_1 *and* φ_k *for* $k \in [2,m]$ *are strictly increasing in* $[-r_{s(1)}, R_*)$ *and* $[-r_{s(k)}, -r_{s(k-1)})$, *respectively.*

By Remark 3.3.1 there exist k_* and ν_* such that $\varphi_{k_*}(\nu_*) = Y$ and

$$\nu_* \in \begin{cases} (-r_{s(1)}, R_*) & \text{for } k_* = 1\,, \\ (-r_{s(k_*)}, -r_{s(k_*-1)}] & \text{for } k_* \in [2,m]\,. \end{cases} \tag{3.11}$$

Theorem 3.3.1 *Let* $x^* = (x_1^*, \ldots, x_n^*), y^* = (y_1^*, \ldots, y_n^*)$ *be such that*

$$x_i^* = \begin{cases} 0 & \text{for } i \in [1, s(k_*)-1]\,, \\ \dfrac{1}{\lambda_i} \ln\left(\dfrac{R_i + r_i}{R_i - \nu_*}\right) & \text{for } i \in [s(k_*), n]\,, \end{cases} \tag{3.12}$$

$$y_i^* = \begin{cases} 0 & \text{for } i \in [1, s(k_*)-1]\,, \\ \dfrac{1/(\lambda_i(R_i - \nu_*)}{\sum_{j=s(k_*)}^{n} (1/(\lambda_j(R_j - \nu_*))}, & \text{for } i \in [s(k_*), n]\,. \end{cases} \tag{3.13}$$

Then x^* *and* y^* *are the optimal strategies of Searcher and Hider, respectively, and* ν_* *is the value of the game.*

Proof By (3.11), it is clear that (3.13) and (3.12) give well defined strategies x^* and y^* of Searcher and Hider. Also, it is easy to see that

$$M(x^*, y) = -\sum_{i=1}^{s(k_*)-1} r_i y_i + \nu_* \sum_{i=s(k_*)}^{n} y_i \quad \text{for any strategy } y. \tag{3.14}$$

By (3.11), minimum in (3.14) reaches at any y such that $y_i = 0$ for $i \in [1, s(k_*) - 1]$. It is clear that (3.12) satisfies this condition. So,

$$M(x^*, y) \geq M(x^*, y^*) = \nu_* \quad \text{for any strategy } y. \tag{3.15}$$

On the other hand we have the following

$$\frac{\partial}{\partial x_i} M(x^*, y^*) = \lambda_i y_i^* (R_i + r_i) \exp(-\lambda_i x_i^*)$$
$$= \begin{cases} 0 & \text{for } i \in [1, s(k_*) - 1], \\ \lambda_i y_i^* (R_i - \nu_*) = \dfrac{1}{\sum_{j=s(k_*)}^{n} 1/(\lambda_i (R_i - \nu_*))} & \text{for } i \in [s(k_*), n]. \end{cases}$$

Hence, by the Kuhn - Tucker Theorem (Mangasarian [71]), $M(x, y^*)$ reaches its maximum at x^* given by (3.12). This fact jointly with (3.15) completes the proof of Theorem 3.3.1.

3.4 Two-Sided Allocation Game Without Search Cost

Danskin [26] and Croucher [25] studied a zero-sum two-sided allocation game. In formulating our game we will present a different scenario from that of Croucher [25]. There are two players, Searcher and Protector, who know only that an object has been hidden at the point i of the integer interval $[1, n]$ with probability σ_i. We note two interpretations of this set-up; a third party may have deposited a valuable object to Searcher at one of n sites or there are n sites each of which may yield a valuable resource to Searcher. In the latter case, for our model to be applicable, we would need to be sure that the probabilities are independent. Searcher undertakes searches by allocating resources x_i to the point i of $[1, n]$ in such a way that $\sum_{i=1}^{n} x_i = X$, where X is the total amount of continuous search effort available to him. In like manner, Protector has a continuous amount of resource Y available to him and, by allocating resource y_i to the point i, where $\sum_{i=1}^{n} y_i = X$, he will make it more difficult for Searcher to find an object that is located at the point i. Following Croucher, we will assume that the probability of Searcher finding an object located at the point i when Searcher uses strategy x and Protector strategy

y is given by $\rho_i'(x,y) = q_i(1-\exp(-\lambda_i x_i))\exp(-\mu_i y_i)$ where $0 < q_i \le 1$. Our expression contains a q_i which Croucher does not have because we want to allow the possibility that Searcher may not find an object located at a point even if Searcher allocates unlimited resources to the point and Protector none. To keep the model general, we will assume that an object located at point i has value v_i to the Searcher. Then the Searcher's payoff $M(x,y)$ is given by

$$M(x,y) = \sum_{i=1}^{n} p_i(1-\exp(-\lambda_i x_i))\exp(-\mu_i y_i),$$

where $p_i = v_i q_i$, $\lambda_i \ge 0$ and $\mu_i \ge 0$.

Remark 3.4.1 *Since M is concave in x for fixed y and $-M$ is concave in y for fixed x, the Kuhn - Tucker Theorem (Mangasarian [71]) implies that (x^*, y^*) is a Nash equilibrium if and only if there are non - negative ω and ν such that*

$$M_{1i}(x^*,y^*) = \frac{\partial}{\partial x_i} M(x^*,y^*) = p_i\lambda_i \exp(-\lambda_i x_i^* - \mu_i y_i^*) \begin{cases} = \omega & \text{for } x_i^* > 0, \\ \le \omega & \text{for } x_i^* = 0 \end{cases} \tag{3.16}$$

and

$$M_{2i}(x^*,y^*) = -\frac{\partial}{\partial y_i} M(x^*,y^*) = p_i\mu_i(1-\exp(-\lambda_i x_i^*))\exp(-\mu_i y_i^*) \begin{cases} = \nu & \text{for } y_i^* > 0, \\ \le \nu & \text{for } y_i^* = 0. \end{cases} \tag{3.17}$$

Remark 3.4.2 $\omega\nu > 0$.

Proof Since there is a $i \in [1,n]$ such thar $x_i^* > 0$ then, by (3.16), $\omega > 0$. Thus, by (3.17), $\nu > 0$ either. Now the result follows.

For positive ω and ν, introduce the following notation.

$$T(\omega,\nu) = \{i \in [1,n] : \lambda_i p_i \le \omega\},$$
$$L(\omega,\nu) = \{i \in [1,n] : \mu_i\omega < \mu_i\lambda_i p_i \le \mu_i\omega + \lambda_i\nu\},$$
$$Q(\omega,\nu) = \{i \in [1,n] : \mu_i\lambda_i p_i > \mu_i\omega + \lambda_i\nu\},$$

$$x_i(\omega,\nu) = \begin{cases} \frac{1}{\lambda_i}\ln\frac{\lambda_i p_i}{\omega} & \text{for } i \in L(\omega,\nu), \\ \frac{1}{\lambda_i}\ln\frac{\mu_i\omega + \lambda_i\nu}{\mu_i\omega} & \text{for } i \in Q(\omega,\nu), \\ 0 & \text{otherwise}, \end{cases}$$

$$y_i(\omega,\nu) = \begin{cases} \frac{1}{\mu_i}\ln\frac{\lambda_i\mu_i p_i}{\mu_i\omega + \lambda_i\nu} & \text{for } i \in Q(\omega,\nu), \\ 0 & \text{otherwise}, \end{cases}$$

$$K(\omega,\nu)=\sum_{i=1}^{n}x_i(\omega,\nu),$$

$$H(\omega,\nu)=\sum_{i=1}^{n}y_i(\omega,\nu).$$

Note that $\lambda_i=0$ implies $i\in T(\omega,\nu)$ for all non-negative ω and ν and that $\mu_i=0$ implies $i\notin Q(\omega,\nu)$ for all non-negative ω and ν. Thus the $x_i(\omega,\nu)$ and $y_i(\omega,\nu)$ are well-defined.

3.4.1 Optimal Strategies of the Game

We first develop some properties of the expressions introduced in the previous section.

Lemma 3.4.1 *(i)For fixed $\omega>0$ and $0<\nu_1<\nu_2$, we have $x_i(\omega,\nu_1)\le x_i(\omega,\nu_2)$ and $y_i(\omega,\nu_1)\ge y_i(\omega,\nu_2)$. Further $K(\omega,\nu_1)\le K(\omega,\nu_2)$ and $H(\omega,\nu_1)\ge H(\omega,\nu_2)$ where, in each case, equality holds if and only if $Q(\omega,\nu_1)=\emptyset$.*

(ii) For fixed $\nu>0$ and $0<\omega_1<\omega_2$, we have $x_i(\omega_1,\nu)\ge x_i(\omega_2,\nu)$ and $y_i(\omega_1,\nu)\ge y_i(\omega_2,\nu)$. Further $K(\omega_1,\nu)\ge K(\omega_2,\nu)$ and $H(\omega_1,\nu)\ge H(\omega_2,\nu)$ where, in the first case, equality holds if and only if $T(\omega_1,\nu)=[1,n]$ and, in the second case, equality holds if and only if $Q(\omega_1,\nu)=\emptyset$.

(iii) For fixed $\omega>0$ and $\nu>0$, $H(\omega,\cdot)$, $K(\omega,\cdot)$, $H(\cdot,\nu)$, and $K(\cdot,\nu)$ are continuous functions.

Proof First note that if

$$\Omega(\omega,\nu)=L(\omega,\nu)\cup Q(\omega,\nu)=\{i\in[1,n]:\lambda_ip_i>\omega\}$$

then
(a) $\Omega(\omega,\nu)$ does not depend on ν,
(b) $\Omega(\omega_2,\nu)\subseteq\Omega(\omega_1,\nu)$ for $0<\omega_1<\omega_2$.

(i) For fixed $\omega>0$ and $0<\nu_1<\nu_2$, we have $L(\omega,\nu_1)\subseteq L(\omega,\nu_2)$ and $Q(\omega,\nu_1)\supseteq Q(\omega,\nu_2)$. So, it is enough to consider separately the cases $i\in T(\omega,\nu_1)$, $i\in L(\omega,\nu_1)$, $i\in Q(\omega,\nu_2)$ and $i\in Q(\omega,\nu_1)\cap L(\omega,\nu_2)$.

If $i\in T(\omega,\nu_1)$, then

$$x_i(\omega,\nu_1)=x_i(\omega,\nu_2)=0.$$

If $i\in L(\omega,\nu_1)$, then

$$x_i(\omega,\nu_1)=x_i(\omega,\nu_2)=\frac{1}{\lambda_i}\ln\frac{\lambda_ip_i}{\omega}.$$

If $i\in Q(\omega,\nu_2)$, then, since $\nu_1<\nu_2$,

$$x_i(\omega,\nu_1) = \frac{1}{\lambda_i}\ln\frac{\mu_i\omega+\lambda_i\nu_1}{\mu_i\omega}$$
$$< \frac{1}{\lambda_i}\ln\frac{\mu_i\omega+\lambda_i\nu_2}{\mu_i\omega} = x_i(\omega,\nu_2)\,.$$

If $i \in L(\omega,\nu_2)\cap Q(\omega,\nu_1)$, then, since $\mu_i\lambda_i p_i > \mu_i\omega+\lambda_i\nu_2$ for $i \in Q(\omega,\nu_2)$ and $\nu_1 < \nu_2$, we have that

$$x_i(\omega,\nu_1) = \frac{1}{\lambda_i}\ln\frac{\mu_i\omega+\lambda_i\nu_1}{\mu_i\omega} < \frac{1}{\lambda_i}\ln\frac{\mu_i\omega+\lambda_i\nu_2}{\mu_i\omega}$$
$$\le \frac{1}{\lambda_i}\ln\frac{\lambda_i p_i}{\omega} = x_i(\omega,\nu_2)\,.$$

Thus, $x_i(\omega,\nu)$ is non-decreasing on ν and strictly increasing on ν while $Q(\omega,\nu)\neq\emptyset$. Therefor, the same can be told of $K(\omega,\nu)$.

Since $Q(\omega,\nu_2)\subseteq Q(\omega,\nu_1)$ for $\nu_2>\nu_1$, the decreasing of $y_i(\omega,\nu)$ and $H(\omega,\nu)$ on ν is clear. Now (i) follows.

The proofs of (ii) and (iii) follow in an analogous manner.

Lemma 3.4.2 *For $i \in [1,n]$ and all positive ω and ν, the conditions for M_{1i} and M_{2i} given in Remark 3.4.1 are satisfied by $x_i(\omega,\nu)$ and $y_i(\omega,\nu)$.*

Proof Straightforward.

Note that for a fixed positive ν

(a) $K(\omega,\nu)\to\infty$ as $\omega\to 0$,

(b) $K(\omega,\nu)=0$ for $\omega\geq\bar{\omega}$ where $\bar{\omega}$ is the minimal positive such that $L(\omega,\nu)\cup Q(\omega,\nu)=\emptyset$,

(c) $K(\omega,\nu)$ is continuous for $\omega>0$, and, by Lemma 3.4.1, strictly decreasing for $\omega\in(0,\bar{\omega})$.

So, for any positive ν there is unique positive $\omega(\nu)$ such that $K(\omega(\nu),\nu)=X$. It is clear that $\omega(\nu)$ is continuous for $\nu>0$. Also, $\omega(\nu)\to 0$ as $\nu\to 0$. Therefore, $H(\omega(\nu),\nu)$ is continuous, and $H(\omega(\nu),\nu)\to\infty$ as $\nu\to 0$ and $H(\omega(\nu),\nu)\to 0$ as $\nu\to\infty$. So, there is a ν_* such that $H(\omega(\nu_*),\nu_*)=Y$.

Theorem 3.4.1 *$x(\omega(\nu_*),\nu_*)$ and $y(\omega(\nu_*),\nu_*)$ are the optimal strategies of Searcher and Protector respectively.*

Proof The result follows from Remark 3.4.1 and Lemma 3.4.2.

3.5 Two-Sided Allocation Game with Search Cost

Consider a variant of the two-sided allocation game where Searcher can apply any amount of search effort from 0 to X but Protector is going to use all his resources Y. Searcher is to maximize his reward minus search cost (which equals to $C\sum_{i=1}^{n} x_i$, where $C>0$). So, the Searcher's payoff $M(x,y)$ is given by

$$M(x,y) = \sum_{i=1}^{n} p_i(1-\exp(-\lambda_i x_i))\exp(-\mu_i y_i) - C\sum_{i=1}^{n} x_i .$$

where $\sum_{i=1}^{n} x_i \le X$, $\sum_{i=1}^{n} y_i = Y$ and $x_i \ge 0$, $y_i \ge 0$ for $i \in [1, n]$.

Remark 3.5.1 *Since M is concave in x for fixed y and $-M$ is concave in y for fixed x, the Kuhn - Tucker Theorem (Mangasarian [71]) implies that (x^*, y^*) is a Nash equilibrium if and only if there are non - negative ω and ν such that*

$$M_{1i}(x^*, y^*) = \frac{\partial}{\partial x_i} M(x^*, y^*) = p_i\lambda_i \exp(-\lambda_i x_i^* - \mu_i y_i^*) - C \begin{cases} = \omega & \text{for } x_i^* > 0 , \\ \le \omega & \text{for } x_i^* = 0 \end{cases}$$

and

$$M_{2i}(x^*, y^*) = -\frac{\partial}{\partial y_i} M(x^*, y^*) = p_i\mu_i(1-\exp(-\lambda_i x_i^*))\exp(-\mu_i y_i^*) \begin{cases} = \nu & \text{for } y_i^* > 0 , \\ \le \nu & \text{for } y_i^* = 0 , \end{cases}$$

where

$$\omega \begin{cases} \ge 0 & \text{for } \sum_{i=1}^{n} x_i^* = X , \\ = 0 & \text{otherwise} . \end{cases}$$

For non-negative ω and ν, let

$$L(\omega, \nu) = \{i \in [1, n] : \mu_i(C+\omega) < \mu_i\lambda_i p_i \le \mu_i(C+\omega) + \lambda_i\nu\} ,$$

$$Q(\omega, \nu) = \{i \in [1, n] : \mu_i\lambda_i p_i > \mu_i(C+\omega) + \lambda_i\nu\} ,$$

$$x_i(\omega, \nu) = \begin{cases} \frac{1}{\lambda_i}\ln\frac{\lambda_i p_i}{C+\omega} & \text{for } i \in L(\omega,\nu) , \\ \frac{1}{\lambda_i}\ln\frac{\mu_i(C+\omega)+\lambda_i\nu}{\mu_i(C+\omega)} & \text{for } i \in Q(\omega,\nu) , \\ 0 & \text{otherwise} , \end{cases}$$

$$y_i(\omega, \nu) = \begin{cases} \frac{1}{\mu_i}\ln\frac{\lambda_i\mu_i p_i}{\mu_i(C+\omega)+\lambda_i\nu} & \text{for } i \in Q(\omega,\nu) , \\ 0 & \text{otherwise} , \end{cases}$$

$$K(\omega,\nu) = \sum_{i=1}^{n} x_i(\omega,\nu), \quad H(\omega,\nu) = \sum_{i=1}^{n} y_i(\omega,\nu) .$$

Lemma 3.5.1 *For $i \in [1, n]$ and all non-negative ω and ν, the conditions for M_{1i} and M_{2i} given in Remark 3.5.1 are satisfied by $x_i(\omega, \nu)$ and $y_i(\omega, \nu)$.*

Proof Straightforward.

Theorem 3.5.1
$$H(0,0) \leq Y \tag{3.18}$$
then $x^* = (0,\ldots,0)$ *and* y^* *are optimal strategies of Searcher and Protector where*

$$y_i^* \begin{cases} \geq y_i(0,0) \text{ such that } \sum_{i=1}^n y_i^* = Y & \text{for } i \in Q(0,0), \\ = 0 & \text{otherwise}. \end{cases}$$

Proof The result follows from Remark 3.5.1 because $M_{2i}(x^*, y^*) \leq C - C = 0$ and $M_{1i}(x^*, y^*) = 0$ for all i.

Clearly $H(\omega, \nu)$ is a continuous function which is non-increasing in ω for fixed ν and non-increasing in ν for fixed ω such that $H(0,\nu) \to 0$ as $\nu \to \infty$. So, if (3.18) does not hold there is unique ν_0 such that $H(0,\nu_0) = Y$.

Theorem 3.5.2 *If (3.18) does not hold and*

$$K(0,\nu_0) \leq X$$

then $x(0,\nu_0)$ *and* $y(0,\nu_0)$ *are optimal strategies of Searcher and Protector.*

Proof Straightforward from Lemma 3.5.1 and Remark 3.5.1.

Since $H(\cdot,\nu)$ is non-increasing for fixed ν, for $\nu \in [0,\nu_0]$, $H(0,\nu) \geq H(0,\nu_0) = Y$. Since $H(\cdot,\nu)$ is continuous and $H(\omega,\nu) \to 0$ as $\omega \to \infty$, there is an $\omega(\nu) \geq 0$ such that $H(\omega(\nu),\nu) = Y$. Clearly $\omega(\cdot)$ is continuous. Then $K(\omega(\nu),\nu)$ is continuous, $K(\omega(\nu_0),\nu_0) = K(0,\nu_0) > X$ and, since $K(\cdot,0)$ is non-increasing function,

$$K(\eta(0),0) = K(\eta(0),0) \leq K(0,0) \leq \text{(by the assumption)} \leq X.$$

Hence there is a $\nu_* \geq 0$ such that $K(\omega(\nu_*),\nu_*) = X$.

Theorem 3.5.3 *If (3.18) does not hold and*

$$K(0,\nu_0) > X$$

then $x(\omega(\nu_*),\nu_*)$ *and* $y(\omega(\nu_*),\nu_*)$ *are optimal strategies of Searcher and Protector.*

Proof Straightforward from Lemma 3.5.1 and Remark 3.5.1.

Theorem 3.5.4 *For fixed* C, λ_i *and* μ_i *where* $i \in [1,n]$, *the game has an optimal strategy.*

Proof The result follows for Theorems 3.5.1 to 3.5.3 cover all possible values of X.

If Searcher and Protector have equivalent facilities, i.e. $\lambda_i = \mu_i$ for $i \in [1, n]$ the optimal strategies can be found without introduction of the auxiliary function $\omega(\cdot)$. Namely, it is clear that

$$K(\omega, \nu) + H(\omega, \nu) = \bar{K}(\omega),$$

where

$$\bar{K}(\omega) = \sum_{\lambda_i p_i > C + \omega} \frac{1}{\lambda_i} \ln \frac{\lambda_i p_i}{C + \omega}.$$

Then, $\bar{K}(\omega)$ is continuous and non-increasing function such that $\bar{K}(\omega) \to 0$ as $\omega \to \infty$. Thus, if $\bar{K}(0) \geq X + Y$ there is unique non-negative ω^* such that $\bar{K}(\omega^*) = X + Y$. Also, $L(\omega, 0) = \emptyset$ so, $H(\omega, 0) = \bar{K}(\omega)$. Then, recalling that for any fixed ω, $H(\omega, \nu)$ is continuous and non-increasing function for ν such that $H(\omega, \nu) \to 0$ as $\nu \to \infty$, we have that if $\bar{K}(0) \geq X + Y$ then there is unique positive ν^* such that $H(\omega^*, \nu^*) = Y$.

Theorem 3.5.5 *Let $\lambda_i = \mu_i$ for $i \in [1, n]$ then*
(i) if $\bar{K}(0) \leq X + Y$ then $x(0, \nu_0)$ and $y(0, \nu_0)$ are optimal strategies of Searcher and Protector,
(ii) if $\bar{K}(0) > X + Y$ then $x(\omega^*, \nu^*)$ and $y(\omega^*, \nu^*)$ are optimal strategies of Searcher and Protector.

Proof Straightforward from Lemma 3.5.1 and Remark 3.5.1.

3.6 A Non-Zero Sum Two-Sided Allocation Game

Bason and Garnaev [20] considered a generalization of the two-sided allocation game for the case where allocating x_i of resource by Searcher to the point i entails a cost of $C_S x_i$ where $C_S > 0$ is a constant, where $\sum_{i=1}^{n} x_i \leq X$, X is the total amount of continuous search effort available to him. In like manner, Protector has a continuous amount of resource Y available to him and, by allocating resource y_i to the point i, he will make it more difficult for the Searcher to find an object that is located at the point i. Again the allocation of resource y_i to the point i involves a cost, this time of $C_P y_i$ where $C_P > 0$. Then the Searcher's payoff $M_1(x, y)$ is given by

$$M_1(x, y) = \sum_{i=1}^{n} p_i (1 - \exp(-\lambda_i x_i)) \exp(-\mu_i y_i) - C_S \sum_{i=1}^{n} x_i.$$

A natural payoff $M_2(x, y)$ for Protector is then

$$M_2(x,y) = -\sum_{i=1}^{n} p_i(1-\exp(-\lambda_i x_i))\exp(-\mu_i y_i) - C_P \sum_{i=1}^{n} y_i.$$

For non-negative ω and ν, let

$$L(\omega,\nu) = \{i \in [1,n] : \mu_i(C_S+\omega) < \mu_i\lambda_i p_i \le \mu_i(C_S+\omega)+\lambda_i(C_P+\nu)\},$$

$$Q(\omega,\nu) = \{i \in [1,n] : \mu_i\lambda_i p_i > \mu_i(C_S+\omega)+\lambda_i(C_P+\nu)\},$$

$$x_i(\omega,\nu) = \begin{cases} \frac{1}{\lambda_i}\ln\frac{\lambda_i p_i}{C_S+\omega} & \text{for } i \in L(\omega,\nu), \\ \frac{1}{\lambda_i}\ln\frac{\mu_i(C_S+\omega)+\lambda_i(C_P+\nu)}{\mu_i(C_S+\omega)} & \text{for } i \in Q(\omega,\nu), \\ 0 & \text{otherwise}, \end{cases}$$

$$y_i(\omega,\nu) = \begin{cases} \frac{1}{\mu_i}\ln\frac{\lambda_i\mu_i p_i}{\mu_i(C_S+\omega)+\lambda_i(C_P+\nu)} & \text{for } i \in Q(\omega,\nu), \\ 0 & \text{otherwise}, \end{cases}$$

$$K(\omega,\nu) = \sum_{i=1}^{n} x_i(\omega,\nu), \qquad H(\omega,\nu) = \sum_{i=1}^{n} y_i(\omega,\nu)$$

and

$$N(\omega,\nu) = (x(\omega,\nu), y(\omega,\nu)).$$

Baston and Garnaev [20] proved the following result.

Theorem 3.6.1 *If $C_S C_P > 0$ then in this game there is unique Nash equilibrium. Also,*
(a) if $K(0,0) \le X$ and $H(0,0) \le Y$, then $N(0,0)$ is a Nash equilibrium,
(b) if $K(0,0) > X$ and $H(\omega_0,0) \le Y$, where ω_0 is unique root of the equation $K(\omega,0) = X$, then $N(\omega_0,0)$ is a Nash equilibrium,
(c) if $H(0,0) > Y$ and $K(0,\nu_0) \le X$, where ν_0 is unique root of the equation $H(0,\nu) = Y$, then $N(0,\nu_0)$ is a Nash equilibrium,
(d) if

$$K(0,0) > X, \quad H(0,0) > Y$$

and

$$K(0,\nu_0) > X, \quad H(\omega_0,0) > Y,$$

then there are positive ω_* and ν_* such that $N(\omega_*,\nu_*)$ is a Nash equilibrium.

This model can have the following interpretation.

A major power A is seeking to increase its ability to attack another power B. To accomplish this, A has n projects in which it can invest resources; although it does not know whether a project i will be successful, it does have an estimate of the probability (q_i in our analysis) of the feasibility of a successful outcome if sufficiently large resources are allocated to it. Furthermore it also has an estimate of the project's value (v_i) if it is successful. Power B knows of A's intention and the projects under consideration by A. It has resources which it can allocate to promote research into countermeasures which will lessen the effect of a successful project.

Thus, in a naive way, our game is a model of the confrontation of the USA and USSR in Star Wars; in this context it is perhaps natural to think of the Searcher as being more aggressive than the Protector. However slightly different viewpoints of Star Wars can lead to the Super Powers being given different roles. The USA can be labelled as Searcher if one regards the USA as trying to develop projects not only to create an effective system of defence and attack but also to exhaust the resources of the USSR. On the other hand the USSR can be labelled as Searcher if one regards the USA's role as primarily defensive and designed to free it from the threat of missile attack. To investigate whether the assignment of roles makes any difference, we will make the reasonable assumptions (certainly in hindsight), that the USA had considerably more resources than the USSR and also, that the costs of allocating resources were relatively much greater for the USSR than the USA.

With the USSR as Protector, we have C_P large and C_S comparatively small; furthermore the values of the p_i are likely to be small as there was a strong body of opinion which gave the projects little chance of success. Hence one might expect $H(0,0)$ to be numerically small and less than the USSR's resources Y. In either case it suggests that the wise course for the USSR is to commit few, if any, resources to the confrontation.

With the USA as Protector, C_S and Y are large whereas C_P is comparatively small. Thus the situation probably gives rise to the case $K(0,0) \leq X$ and $H(0,0) \leq Y$. As before, we conclude that it would have been wise for the USSR not to have entered the confrontation or, at any rate, to have spent comparatively little on it.

The model with respect to the USSR seems robust as different interpretations have given rise to the same conclusion. However the position concerning the USA is much less clear. The USA clearly won the Star Wars battle but it would require a much deeper analysis to determine whether the USA acted in a near optimal way.

3.7 One Person Search Game

Consider the following one person search game. An object has been hidden at the point i of the integer interval $[1,n]$ with probability p_i. Searcher undertakes search by allocating resources x_i to the point i of $[1,n]$ in such a way that $\sum_{i=1}^{n} x_i \le X$, where X is the total amount of continuous search effort available to him. The probability of Searcher finding an object located at the point i when Searcher uses strategy x is given by $p_i(1-\exp(-\lambda_i x_i))$. Reward for detecting the object is R and search at the point i entails cost C_i. Then, the Searcher's payoff $M(x)$ is given by

$$M(x) = \sum_{i=1}^{n} Rp_i(1-\exp(-\lambda_i x_i)) - \sum_{i=1}^{n} C_i x_i .$$

Remark 3.7.1 *Since M is concave in x, the Kuhn - Tucker Theorem (Mangasarian [71]) implies that x^* is optimal strategy if and only if there is a non - negative ω*

$$\frac{\partial}{\partial x_i} M(x^*) = Rp_i\lambda_i \exp(-\lambda_i x_i^*) - C_i \begin{cases} = \omega & \text{for } x_i^* > 0, \\ \le \omega & \text{for } x_i^* = 0, \end{cases}$$

where

$$\omega \begin{cases} \ge 0 & \text{for } \sum_{i=1}^{n} x_i^* = X, \\ = 0 & \text{otherwise}. \end{cases}$$

If $Rp_i\lambda_i \le C_i$ for $i \in [1,n]$ then, by Remark 3.7.1, the strategy $x^* = (0,\dots,0)$ is optimal. So, we assume that there is an i such that $Rp_i\lambda_i > C_i$. Let

$$C^* = \min\{C_i : Rp_i\lambda_i > C_i\} .$$

For non-negarive ω, let

$$Q(\omega) = \{i : Rp_i\lambda_i - C_i > \omega\},$$

$$x_i(\omega) = \begin{cases} \dfrac{1}{\lambda_i} \ln \dfrac{Rp_i\lambda_i}{C_i+\omega} & \text{if } i \in Q(\omega), \\ 0 & \text{otherwise} \end{cases}$$

and

$$K(\omega) = \sum_{i=1}^{n} x_i(\omega) .$$

Clearly K is a continuous function in $[C^*,\infty)$ for $C^* > 0$ and in $(0,\infty)$ for $C^* = 0$ satisfying $K(\omega) \to 0$ as $\omega \to \infty$ for all C^* and $K(\omega) \to \infty$ as $\omega \to 0$ for $C^* = 0$. Thus, if $K(0) > X$, there is a positive $\bar{\omega}$ such that $K(\bar{\omega}) = X$; it is easy to check that the $\bar{\omega}$ is unique.

Theorem 3.7.1 *The strategy $x^*(\omega^*)$ where*

$$\omega^* = \begin{cases} 0 & \text{if } K(0) \le X \text{ and } C^* > 0\,, \\ \bar{\omega} & \text{where } K(\bar{\omega}) = X \text{ if } K(0) > X \text{ or } C^* = 0 \end{cases}$$

is optimal.

Proof Straightforward from Remark 3.7.1.

If $C_i = 0$ for $i \in [1, n]$ then the next corollary shows that the optimal strategy can be expressed easier in terms of λ_i, p_i, R and X.

Without loss of generality we can assume that

$$p_1\lambda_1 \le p_2\lambda_2 \le \ldots p_n\lambda_n\,.$$

Corollary 3.7.1 *If $C_i = 0$ for $i \in [1, n]$ then x^* is the optimal strategy of Searcher where*

$$x_i^* = \begin{cases} \frac{1}{\lambda_i} \ln \frac{Rp_i\lambda_i}{\omega_*} & \text{if } i \in [k, n]\,, \\ 0 & \text{otherwise}\,, \end{cases}$$

$$\omega_* = \exp\left(\frac{\sum\limits_{i=k}^{n} \ln(Rp_i\lambda_i)/\lambda_i - X}{\sum\limits_{i=k}^{n} (1/\lambda_i)} \right)$$

and $k \in [1, n]$ such that

$$\begin{aligned} k = 1 \quad &\text{if} \quad \varphi_1 < X\,, \\ k \in [2, n] \quad &\text{if} \quad \varphi_k < X \le \varphi_{k-1} \end{aligned}$$

and

$$\varphi_s = \sum_{i=s}^{n} \ln(\lambda_i p_i/(\lambda_s p_s))/\lambda_i\,.$$

Proof It is clear that φ_s is decreasing for $s \in [1, n-1]$ such that $\varphi_n = 0$. Now the result follows straightforward from Remark 3.7.1 and Theorem 3.7.1.

3.8 Marketing Games Without Taking into Acount Expenses

Friedman [33] considered the following non-zero sum two-sided allocation game in marketing. Two firms, say 1 and 2, compete against each other in n independent markets. The total sales potential in each market is fixed and is shared between the competitors on the basis of both the quality and the magnitude of the marketing efforts expended by each competitor. Let V_i denote the sales potential of the ith market. If firm 1 and firm 2 allocates x_i and y_i dollars of effort to market i, respectively. Then the share of market i getting by firm 1 and 2 is $V_i x_i/(x_i + y_i)$ and $V_i y_i/(x_i + y_i)$, respectively. Also, budget of firm 1 and 2 is X and Y, respectively. The aim of each firm is to plan its budget so that to maximize total firm profit. So, the payoffs to firms if firm 1 and 2 employ strategy x and y, respectively, are given by

$$M_1(x, y) = \sum_{i=1}^{n} \frac{V_i x_i}{x_i + y_i},$$
$$M_2(x, y) = \sum_{i=1}^{n} \frac{V_i y_i}{x_i + y_i},$$

where $\sum_{i=1}^{n} x_i = X, \sum_{i=1}^{n} y_i = Y$ and $x_i \geq 0, y_i \geq 0$ for $i \in [1, n]$.

Remark 3.8.1 *Since M_1 and M_2 is concave in x and y respectively, the Kuhn - Tucker Theorem (Mangasarian [71]) implies that (x^*, y^*) is a Nash equilibrium if and only if there are non - negative ω and ν such that*

$$\begin{aligned} \frac{\partial}{\partial x_i} M(x^*, y^*) &= \frac{V_i y_i^*}{(x_i^* + y_i^*)^2} \begin{cases} = \omega & \text{for } x_i^* > 0, \\ \leq \omega & \text{for } x_i^* = 0, \end{cases} \\ \frac{\partial}{\partial y_i} M(x^*, y^*) &= \frac{V_i x_i^*}{(x_i^* + y_i^*)^2} \begin{cases} = \nu & \text{for } y_i^* > 0, \\ \leq \nu & \text{for } y_i^* = 0. \end{cases} \end{aligned} \tag{3.19}$$

Theorem 3.8.1 *There is a unique Nash equilibrium (x^*, y^*) with the payoff vector $(VX/(X+Y), VY/(X+Y))$ and*

$$x_i^* = \frac{V_i X}{V},$$
$$y_i^* = \frac{V_i Y}{V},$$

where $i \in [1, n], V = \sum_{i=1}^{n} V_i$.

Proof Let (x^*, y^*) be a Nash equilibrium. Then, by Remark 3.8.1, there is non-negative ω and ν such that (3.19) holds.

Assume that there is a i such that $x_i^* y_i^* = 0$. Say, $x_i^* = 0$. Then, by (3.19), $y_i^* = 0$. This contradiction shows that $x_i^* > 0$ and $y_i^* > 0$ for $i \in [1, n]$. So, by Remark 3.8.1, $\omega\nu > 0$. Thus, (3.19) implies that

$$\omega x_i^* = \nu y_i^*,$$
$$x_i^* + y_i^* = \frac{V_i}{\omega + \nu}$$

and

$$x_i^* = \frac{\nu V_i}{(\omega + \nu)^2},$$
$$y_i^* = \frac{\omega V_i}{(\omega + \nu)^2}.$$

Summing up the first two equations on i yields that

$$\omega X = \nu Y,$$
$$X + Y = \frac{V}{\omega + \nu}.$$

So,

$$\omega = \frac{YV}{(X+Y)^2},$$
$$\nu = \frac{XV}{(X+Y)^2}.$$

Thus, x^* and y^* are to be given by (3.8.1). This completes the proof of Theorem 3.8.1.

Following Monahan [73] consider a variant of the Friedman allocation game where the payoffs to firms if firm 1 and 2 employ strategy x and y, respectively, are given by

$$M_1(x,y) = \sum_{i=1}^{n} \frac{V_i \alpha_i x_i}{\alpha_i x_i + \beta_i y_i},$$
$$M_2(x,y) = \sum_{i=1}^{n} \frac{V_i \beta_i y_i}{\alpha_i x_i + \beta_i y_i},$$

where α_i and β_i are positive for $i \in [1, n]$.

Similarly to Theorem 3.8.1 the following result can be proved.

Theorem 3.8.2 *There is a unique Nash equilibrium (x^*, y^*) with the payoff vector (η_1, η_2). Also,*

$$x_i^* = \frac{V_i \alpha_i \beta_i \nu}{(\alpha_i \nu + \beta_i \omega)^2},$$
$$y_i^* = \frac{V_i \alpha_i \beta_i \omega}{(\alpha_i \nu + \beta_i \omega)^2} \quad \textit{for} \quad i \in [1, n]$$

and

$$\eta_1 = \sum_{i=1}^{n} \frac{V_i \alpha_i \nu}{\alpha_i \nu + \beta_i \omega},$$
$$\nu = \sum_{i=1}^{n} \frac{V_i \beta_i \omega}{\alpha_i \nu + \beta_i \omega},$$

where

$$\omega = \sum_{i=1}^{n} \frac{V_i \alpha_i \beta_i Y}{(\alpha_i X + \beta_i X)^2},$$
$$\nu = \sum_{i=1}^{n} \frac{V_i \alpha_i \beta_i X}{(\alpha_i X + \beta_i Y)^2}.$$

Let firm 2 allocated a_i dollars on market i and these values are fixed and known to firm 1. Firm 1 and firm 2 are going to invest X and Y dollars respectively on n markets. If firm 1 and firm 2 allocate x_i and y_i dollars on market i then their payoffs are given by

$$M_1(x,y) = \sum_{i=1}^{n} \frac{V_i x_i}{x_i + y_i + a_i},$$
$$M_2(x,y) = \sum_{i=1}^{n} \frac{V_i (y_i + a_i)}{x_i + y_i + a_i},$$

where $\sum_{i=1}^{n} x_i = X, \sum_{i=1}^{n} y_i = Y$ and $x_i \geq 0, y_i \geq 0$ for $i \in [1, n]$.

We shall prove that the game has unique Nash equilibrium.

Remark 3.8.2 *By the Kuhn-Tucker Theorem (Mangasarian [71]), (x^*, y^*) is a Nash equilibrium if and only if there are non-negative ω and ν such that*

$$\frac{\partial}{\partial x_i} M_1(x^*, y^*) = \frac{V_i(a_i + y_i^*)}{(x_i^* + y_i^* + a_i)^2} \begin{cases} = \omega & \text{for } x_i^* > 0, \\ \leq \omega & \text{for } x_i^* = 0 \end{cases} \tag{3.20}$$

and

$$\frac{\partial}{\partial y_i} M_2(x^*, y^*) = \frac{V_i x_i}{(x_i^* + y_i^* + a_i)^2} \begin{cases} = \nu & \text{for } y_i^* > 0, \\ \leq \nu & \text{for } y_i^* = 0. \end{cases} \tag{3.21}$$

Remark 3.8.3 $\omega\nu > 0$.

Proof Since there is a $i \in [1, n]$ such that $y_i^* > 0$ then, by (3.20), $\omega > 0$. Similarly, by (3.21), $\nu > 0$. Now the result follows.

For positive ω and ν, let

$$L(\omega,\nu) = \{i : 1/\omega \leq a_i/V_i\}, \qquad Q(\omega,\nu) = \{i : \omega/(\omega+\nu)^2 \leq a_i/V_i < 1/\omega\},$$
$$M(\omega,\nu) = \{i : a_i/V_i < \omega/(\omega+\nu)^2\},$$

$$x_i(\omega,\nu) = \begin{cases} 0 & \text{if } i \in L(\omega,\nu), \\ \sqrt{V_i a_i/\omega} - a_i & \text{if } i \in Q(\omega,\nu), \\ \nu V_i/(\omega+\nu)^2 & \text{if } i \in M(\omega,\nu), \end{cases}$$
$$y_i(\omega,\nu) = \begin{cases} 0 & \text{if } i \notin M(\omega,\nu), \\ \omega V_i/(\omega+\nu)^2 - a_i & \text{if } i \in M(\omega,\nu), \end{cases}$$

$$K(\omega,\nu) = \sum_{i=1}^{n} x_i(\omega,\nu), \qquad H(\omega,\nu) = \sum_{i=1}^{n} y_i(\omega,\nu)$$

and

$$N(\omega,\nu) = (x(\omega,\nu), y(\omega,\nu)).$$

Lemma 3.8.1 *For $i \in [1, n]$ and all positive ω and ν, the conditions for the partial derivatives in Remark 3.8.3 are satisfied by $x_i(\omega, \nu)$ and $y_i(\omega, \nu)$.*

Proof Straightforward.

Lemma 3.8.2 *Every Nash equilibrium is of the form $N(\omega, \nu)$ for some positive ω and ν.*

Proof Suppose (x^*, y^*) is a Nash equilibrium, then there are non-negative ω and ν such that the conditions of Remark 3.8.2 are satisfied.

If $x_i^* = 0$, then, by (3.21), $y_i^* = 0$ so that (3.20) implies $1/\omega \leq a_i/V_i$. Thus $i \in L(\omega, \nu)$ and $x_i^* = x_i(\omega, \nu)$ and $y_i^* = y_i(\omega, \nu)$.

If $x_i^* > 0$ and $y_i^* = 0$ then, by (3.20), $x_i^* = \sqrt{V_i a_i/\omega} - a_i$ so that $i \notin L(\omega, \nu)$ and (3.21) gives

$$\nu \geq V_i x_i^* / (V_i a_i/\omega) = \sqrt{V_i \omega / a_i} - \omega .$$

Hence $i \in Q(\omega, \nu)$ and $x_i^* = x_i(\omega, \nu)$ and $y_i^* = y_i(\omega, \nu)$.

If $x_i^* > 0$ and $y_i^* > 0$ then, solving (3.20) and (3.21), we see that $x_i^* = x_i(\omega, \nu)$ and $y_i^* = y_i(\omega, \nu)$.

Theorem 3.8.3 *There are unique positive ω_* and ν_* such that $N(\omega_*, \nu_*)$ is Nash equilibrium.*

Proof Let

$$\bar{H}(t) = \sum_{i \in \bar{M}(t)} (tV_i - a_i) ,$$

where

$$\bar{M}(t) = \{i : t > a_i/V_i\} .$$

then

$$H(\omega, \nu) = \bar{H}\left(\frac{\omega}{(\omega + \nu)^2}\right) .$$

Since $\bar{H}(0) = 0$, $\bar{H}(t) \to \infty$ as $t \to \infty$ and $\bar{H}(t)$ is non-decreasing and continuous for $t \geq 0$, there is unique positive t_* such that $\bar{H}(t_*) = Y$.
For $\omega \in (0, 1/t_*]$, there is unique $\nu(\omega)$ such that

$$\frac{\omega}{(\omega + \nu(\omega))^2} = t_* . \tag{3.22}$$

Therefore,

$$\nu(\omega) = \sqrt{\omega/t_*} - \omega .$$

It is clear that

(a) $\nu(\omega) \to 0$ as $\omega \to 0$,
(b) $\nu(1/t_*) = 0$.

Let

$$P(\omega) = K(\omega, \nu(\omega)) + H(\omega, \nu(\omega)).$$

Then, by (3.22)

$$\begin{aligned} P(\omega) &= \sum_{t_* \le a_i/V_i < 1/\omega} \left(\sqrt{V_i a_i/\omega} - a_i\right) + \sum_{a_i/V_i < t_*} (V_i/(\omega+\nu) - a_i) \\ &= \sum_{t_* \le a_i/V_i < 1/\omega} \left(\sqrt{V_i a_i/\omega} - a_i\right) + \sqrt{t_*/\omega} \sum_{a_i/V_i < t_*} V_i - \sum_{a_i/V_i < t_*} a_i . \end{aligned}$$

So, $P(\omega)$ is non-increasing and continuous for $\omega > 0$ and strictly decreasing while $P(\omega) > 0$. Also, $P(1/t_*) = \bar{H}(t_*) = Y < X + Y$ and $P(\omega) \to \infty$ as $\omega \to 0$. Thus, there is unique $\omega_* > 0$ such that

$$P(\omega_*) = X + Y .$$

So, for $H(\omega_*, \nu(\omega_*)) = Y$,

$$K(\omega_*, \nu(\omega_*)) = X .$$

Then, by Lemma 3.8.2, $N(\omega_*, \nu(\omega_*))$ is a Nash equilibrium. Now the result follows.

3.9 Marketing Games Taking into Acount Expenses

Consider a generalization of the Friedman allocation game where firms are to miximize their profits minus expences. So, the payoffs to firms if firm 1 and 2 employ strategy x and y, respectively, are given by

$$M_1(x, y) = \sum_{i=1}^{n} \frac{V_i x_i}{x_i + y_i} - C_1 \sum_{i=1}^{n} x_i ,$$

$$M_2(x, y) = \sum_{i=1}^{n} \frac{V_i y_i}{x_i + y_i} - C_2 \sum_{i=1}^{n} y_i ,$$

where C_2 and C_2 are positive constant describing expenses and $\sum_{i=1}^{n} x_i \le X, \sum_{i=1}^{n} y_i \le Y$ and $x_i \ge 0, y_i \ge 0$ for $i \in [1, n]$.

Remark 3.9.1 *By the Kuhn-Tucker Theorem [71]), (x^*, y^*) is a Nash equilibrium if and only if there are non-negative ω and ν such that*

$$\frac{\partial}{\partial x_i} M_1(x^*, y^*) = \frac{V_i y_i^*}{(x_i^* + y_i^*)^2} - C_1 \begin{cases} = \omega & \text{for } x_i^* > 0, \\ \le \omega & \text{for } x_i^* = 0, \end{cases}$$

$$\frac{\partial}{\partial y_i} M_2(x^*, y^*) = \frac{V_i x_i^*}{(x_i^* + y_i^*)^2} - C_2 \begin{cases} = \nu & \text{for } y_i^* > 0, \\ \le \nu & \text{for } y_i^* = 0, \end{cases}$$

where

$$\omega \begin{cases} \geq 0 & \textit{for } \sum_{i=1}^{n} x_i^* = X\,, \\ = 0 & \textit{otherwise}\,, \end{cases}$$
$$\nu \begin{cases} \geq 0 & \textit{for } \sum_{i=1}^{n} y_i^* = Y\,, \\ = 0 & \textit{otherwise}\,. \end{cases}$$

Lemma 3.9.1 *Let (x^*, y^*) be a Nash equilibrium then $x_i^* y_i^* > 0$ for $i \in [1, n]$.*

Proof Assume that there is a i such that $x_i^* y_i^* = 0$. Say, $x_i^* = 0$. Then, by Remark 3.9.1, $y_i^* = 0$. This contradiction shows that $x_i^* > 0$ and $y_i^* > 0$ for $i \in [1, n]$.

Let

$$x_i(\omega, \nu) = V_i \frac{C_2 + \nu}{(C_1 + \omega + C_2 + \nu)^2}\,,$$
$$y_i(\omega, \nu) = V_i \frac{C_1 + \omega}{(C_1 + \omega + C_2 + \nu)^2} \quad \text{for} \quad i \in [1, n]\,,$$

$$K(\omega, \nu) = \sum_{i=1}^{n} x_i(\omega, \nu), \quad H(\omega, \nu) = \sum_{i=1}^{n} y_i(\omega, \nu)$$

and

$$N(\omega, \nu) = (x(\omega, \nu), y(\omega, \nu))\,.$$

Lemma 3.9.2 *For $i \in [1, n]$ and all positive ω and ν, the conditions for the partial derivatives in Remark 3.9.1 are satisfied by $x_i(\omega, \nu)$ and $y_i(\omega, \nu)$.*

Proof Straightforward.

Lemma 3.9.3 *Every Nash equilibrium is of the form $N(\omega, \nu)$ for some non-negative ω and ν.*

Proof Straightforward from Remark 3.9.1 and Lemmas 3.9.1 and 3.9.2.

Lemma 3.9.4 *Let $N(\omega, \nu)$ be a Nash equilibrium.*
(a) If $\omega = 0 = \nu$, then

$$\frac{VC_2}{(C_1 + C_2)^2} \leq X \quad \textit{and} \quad \frac{VC_1}{(C_1 + C_2)^2} \leq Y\,. \tag{3.23}$$

(b) If $\omega = 0$ and $\nu > 0$, then

$$\frac{VY}{(X + Y)^2} \leq C_1 \quad \textit{and} \quad \frac{VC_1}{(C_1 + C_2)^2} > Y\,. \tag{3.24}$$

(c) If $\omega > 0$ and $\nu = 0$, then

$$\frac{VX}{(X + Y)^2} \leq C_2 \quad \textit{and} \quad \frac{VC_2}{(C_1 + C_2)^2} > X\,. \tag{3.25}$$

(d) If $\omega > 0$ and $\nu > 0$, then

$$\frac{VX}{(X + Y)^2} > C_2 \quad \textit{and} \quad \frac{VY}{(X + Y)^2} > C_1\,. \tag{3.26}$$

Proof Let $N(\omega,\nu)$ be a Nash equilibrium.
(a) If $\omega=\nu=0$ then, by Remark 3.9.1,

$$K(0,0)\leq X \quad \text{and} \quad H(0,0)\leq Y.$$

Thus, (3.23) holds.
(b) If $\omega=0$ and $\nu>0$ then, by Remark 3.9.1,

$$K(0,\nu)\leq X \quad \text{and} \quad H(0,\nu)=Y.$$

Thus,

$$V\frac{C_2+\nu}{(C_1+C_2+\nu)^2}\leq X \quad \text{and} \quad V\frac{C_1}{(C_1+C_2+\nu)^2}=Y. \tag{3.27}$$

So, for $\nu>0$, then

$$\frac{VC_1}{(C_1+C_2)^2}>Y$$

and

$$\nu=\sqrt{C_1/Y}-C_1-C_2.$$

Substituting the ν into the inequality of (3.27) implies that

$$\frac{VY}{(X+Y)^2}\leq C_1.$$

Now (3.24) follows.
(c) can be proved in analogous manner.
(d) If $\omega>0$ and $\nu>0$ then, by Remark 3.9.1,

$$K(\omega,\nu)=X \quad \text{and} \quad H(\omega,\nu)=Y.$$

Thus,

$$V\frac{C_2+\nu}{(C_1+\omega+C_2+\nu)^2}=X \quad \text{and} \quad V\frac{C_1+\omega}{(C_1+\omega+C_2+\nu)^2}=Y.$$

So,

$$\omega+C_1=\frac{YV}{(X+Y)^2},$$
$$\nu+C_2=\frac{XV}{(X+Y)^2}.$$

Since ω and ν are positive, the result now follows.

Theorem 3.9.1 *There is a unique Nash equilibrium $N(\omega,\nu)$ with the payoff vector (η_1,η_2). Also,*
(a) if (3.23) holds, then $\omega=0=\nu$ and

$$\eta_1 = V\left(\frac{C_2}{C_1+C_2}\right)^2,$$
$$\eta_2 = V\left(\frac{C_1}{C_1+C_2}\right)^2,$$

(b) if (3.24) holds, then

$$\omega = 0, \quad \nu = \sqrt{C_1/Y} - C_1 - C_2$$

and

$$\eta_1 = \frac{VY}{C_1}\left(\sqrt{C_1/Y} - C_1\right)^2,$$
$$\eta_2 = V\sqrt{Y/C_1} - C_2Y,$$

(c) if (3.25) holds, then

$$\nu = 0, \quad \omega = \sqrt{C_2/X} - C_1 - C_2,$$

and

$$\eta_1 = V\sqrt{X/C_2} - C_1X,$$
$$\eta_2 = \frac{VX}{C_2}\left(\sqrt{C_2/X} - C_2\right)^2,$$

(d) if (3.26) holds, then

$$\omega = \frac{YV}{(X+Y)^2} - C_1,$$
$$\nu = \frac{XV}{(X+Y)^2} - C_2$$

and

$$\eta_1 = \frac{YV}{X+Y} - C_1X,$$
$$\eta_2 = \frac{XV}{X+Y} - C_2Y.$$

Proof Note that inequalities from (3.23) to (3.26) cover all relations between X, Y, C_1 and C_2. Also, they exclude each others. Thus, the result follows from Remark 3.9.1 and Lemma 3.9.3.

When $C_1 = C_2$ the next corollary shows that a Nash equilibrium can be expressed very simply in terms of X, Y, C and V.

Corollary 3.9.1 *Let $C_1 = C_2 = C$ fand $X \leq Y$, then there is a unique Nash equilibrium (x^*, y^*) with the payoff vector (η_1, η_2) where*

$$x_i^* = \begin{cases} V/(4C) & \text{if } X \geq V/(4C), \\ XV_i/V & \text{if } X < V/(4C), \end{cases}$$
$$y_i^* = \begin{cases} V/(4C) & \text{if } X \geq V/(4C), \\ Y_*V_i/V & \text{if } X < V/(4C) \end{cases}$$

and

$$\eta_1 = \eta_2 = V/4 \quad \text{if } X \geq V/(4C),$$
$$\eta_1 = VX/(X+Y_*) - CX, \quad \eta_2 = VY_*/(X+Y_*) - CY_* \quad \text{if } X < V/(4C),$$

where $Y_* = \min\{\sqrt{XV/C} - X, Y\}$.

Problems

Solve the following variant of the search allocation game. Let an object is hidden in one on n boxes. The probability that the object is located in box i is p_i. The value of the object is V. Two players, say, 1 and 2, allocating their continuous resources X and Y respectively, try to find the object. Search of the objects entails cost C_i for player i where $i = 1, 2$. The player, detected the object, receives it, if they both find it they share reward. So, the payoffs to players are given by

$$M_1(x,y) = \sum_{i=1}^{n} Vp_i(1-\exp(-\lambda_i x_i))(1+\exp(-\mu_i y_i))/2 - C_1\sum_{i=1}^{n} x_i\,,$$
$$M_2(x,y) = \sum_{i=1}^{n} Vp_i(1-\exp(-\mu_i y_i))(1+\exp(-\lambda_i x_i))/2 - C_2\sum_{i=1}^{n} y_i\,,$$

where $\sum_{i=1}^{n} x_i \le X$, $\sum_{i=1}^{n} y_i \le Y$ and $x_i \ge 0$, $y_i \ge 0$ for $i \in [1, n]$.

Solve the following three variants of the marketing game.

The first variant of the marketing game. Let firm 2 allocated a_i dollars on market i and these values are fixed and known to firm 1. Firm 1 and firm 2 are going to invest X and Y dollars respectively on n markets. We assume that the firm invested more money on a market overtake it. So, if firm 1 and firm 2 allocate x_i and y_i dollars on market i then their payoffs are given by

$$M_1(x,y) = \sum_{i=1}^{n} V_i|x_i - y_i - a_i|\,,$$
$$M_2(x,y) = \sum_{i=1}^{n} V_i|x_i - y_i - a_i|\,,$$

where $\sum_{i=1}^{n} x_i = X, \sum_{i=1}^{n} y_i = Y$ and $x_i \ge 0, y_i \ge 0$ for $i \in [1, n]$.

The second variant of the marketing game. Let firm 2 allocated a_i dollars on market i and these values are fixed and known to firm 1. Firm 1 and firm 2 are going to invest X and Y dollars respectively on n markets. If firm 1 and firm 2 allocate x_i and y_i dollars on market i then their payoffs are given by

$$M_1(x,y) = \sum_{i=1}^{n} \frac{V_i x_i}{x_i + y_i + a_i} - C_1\sum_{i=1}^{n} x_i\,,$$
$$M_2(x,y) = \sum_{i=1}^{n} \frac{V_i(y_i + a_i)}{x_i + y_i + a_i} - C_2\sum_{i=1}^{n} y_i\,,$$

where C_1, C_2 are non-negative and $\sum_{i=1}^{n} x_i \le X, \sum_{i=1}^{n} y_i \le Y$ and $x_i \ge 0, y_i \ge 0$ for $i \in [1, n]$.

The third variant of the marketing game. Assume that the firm invested more money on a market overtake it. So, if firm 1 and firm 2 allocate x_i and y_i dollars on market i then their payoffs are given by

$$M_1(x,y) = \sum_{i=1}^{n} V_i|x_i - y_i - a_i| - C_1 \sum_{i=1}^{n} x_i ,$$
$$M_2(x,y) = \sum_{i=1}^{n} V_i|x_i - y_i - a_i| - C_2 \sum_{i=1}^{n} y_i .$$

where C_1, C_2 are non-negative and $\sum_{i=1}^{n} x_i \leq X, \sum_{i=1}^{n} y_i \leq Y$ and $x_i \geq 0, y_i \geq 0$ for $i \in [1, n]$.

4 Dynamic Infiltration and Inspection Games

4.1 A Multi - Stage Infiltration Game with a Bunker

Consider the following two - person, zero - sum, multi - stage game. There are two players: Guard and Infiltrator. The Infiltrator's movement is constrained to only integer points: $0, 1, 2, \ldots$ on the non - negative x - axis. If Infiltrator is at point i, he may, one unit of time later, move to point $i-1$, remain at point i or move to point $i+1$. Guard, having a gun with k shots, can shoot at most one shot per unit of time at Infiltrator. Infiltrator has a bunker at the origin 0, where he is immune to the Guard's shooting. It is assumed that Infiltrator knows the number of shots that Guard possesses at all time but Infiltrator does not know where Guard aims his hit and it takes the shot one unit of time to reach x - axis. There is no aiming errors, so Guard can hit any point he desires. If Guard hits the point where Infiltrator locates then the probability of hitting Infiltrator is α where $\alpha \in (0,1)$ and 0 otherwise. Therefore, if Guard observes that Infiltrator is at point i, and if Guard desires to shoot at that instant, he should aim at one of the three points $i-1, i$ and $i+1$. The payoff to Guard is 1 if he hits Infiltrator and 0 otherwise.

Let $\Gamma(i,k)$ denote the game when Infiltrator starts from the point i, and Guard possess k shots. Let $v(i,k)$ be the value of the game $\Gamma(i,k)$. Note that if Infiltrator is at point 1, clearly, he should use the optimal strategy saying him to move at the bunker, then, as initial conditions, we have

$$v(i,0) = 0, i = 1,2,\ldots, \quad v(1,k) = 0, k = 1,2,\ldots. \tag{4.1}$$

On the first step of the game $\Gamma(i,k)$ Infiltrator has three pure strategies

- to move at $i-1$,
- to remain at i,
- to move at $i+1$.

Guard has four strategies (actually he has more than four strategies but only the four bellow are reasonable):

- to shoot at $i-1$,
- to shoot at i,
- to shoot at $i+1$,
- to not shoot at all.

Then the payoff matrix to Guard in $\Gamma(i,k)$ is given by

$$\begin{pmatrix} \alpha+\bar{\alpha}v(i-1,k-1) & v(i,k-1) & v(i+1,k-1) \\ v(i-1,k-1) & \alpha+\bar{\alpha}v(i,k-1) & v(i+1,k-1) \\ v(i-1,k-1) & v(i,k-1) & \alpha+\bar{\alpha}v(i+1,k-1) \\ v(i-1,k) & v(i,k) & v(i+1,k) \end{pmatrix},$$

where $\bar{\xi}=1-\xi$ and Guard picks a row and Infiltrator picks a column. The value of this game is, of course, $v(i,k)$.

Let $x_p(i,k)(y_q(i,k))$ denote the probability that Guard (Infiltrator) emploies his p - st (q - st) strategy, respectively, on the first step of the game $\Gamma(i,k)$. Then $v(i,k)$ is the value of the game $\Gamma(i,k)$ and probability vectors

$$x(i,k)=(x_1(i,k),\ldots,x_4(i,k)),\quad y(i,k)=(y_1(i,k),\ldots,y_3(i,k))$$

are the optimal strategies of Guard and Infiltrator, respectively, if and only if they satisfy the following relations:

$$\begin{aligned}
&\big(\alpha+\bar{\alpha}v(i-1,k-1)\big)y_1 \\
&\qquad +v(i,k-1)y_2+v(i+1,k-1)y_3 && \le v(i,k),\\
&v(i-1,k-1)y_1+\big(\alpha+\bar{\alpha}v(i,k-1)\big)y_2 \\
&\qquad +v(i+1,k-1)y_3 && \le v(i,k),\\
&v(i-1,k-1)y_1+v(i,k-1)y_2 \\
&\qquad +\big(\alpha+\bar{\alpha}v(i+1,k-1)\big)y_3 && \le v(i,k),\\
&v(i-1,k)y_1+v(i,k)y_2+v(i+1,k)y_3 && \le v(i,k),\\
&\big(\alpha+\bar{\alpha}v(i-1,k-1)\big)x_1+v(i-1,k-1)x_2 \\
&\qquad +v(i-1,k-1)x_3+v(i-1,k)x_4 && \ge v(i,k),\\
&v(i-1,k-1)x_1+\big(\alpha+\bar{\alpha}v(i,k-1)\big)x_2 \\
&\qquad +v(i,k-1)x_3+v(i,k)x_4 && \ge v(i,k),\\
&v(i+1,k-1)x_1+v(i+1,k-1)x_2 \\
&\qquad +\big(\alpha+\bar{\alpha}v(i+1,k-1)\big)x_3+v(i+1,k)x_4 && \ge v(i,k).
\end{aligned} \tag{4.2}$$

Here and below for convenience we omit i and k in the notation of vectors $x(i,k), y(i,k)$ and their components. Let

$$V(i,k)=1/(1-v(i,k)),\quad \lambda=\bar{\alpha}.$$

Hence,

$$v(i,k)=1-1/V(i,k). \tag{4.3}$$

Then (4.2) can be rewriten as follows

$$\frac{\lambda x_1+x_2+x_3}{V(i-1,k-1)}+\frac{x_4}{V(i-1,k)}\le\frac{1}{V(i,k)}, \tag{4.4}$$

$$\frac{x_1 + \lambda x_2 + x_3}{V(i,k-1)} + \frac{x_4}{V(i,k)} \le \frac{1}{V(i,k)}, \tag{4.5}$$

$$\frac{x_1 + x_2 + \lambda x_3}{V(i+1,k-1)} + \frac{x_4}{V(i+1,k)} \le \frac{1}{V(i,k)}, \tag{4.6}$$

$$\frac{\lambda y_1}{V(i-1,k-1)} + \frac{y_2}{V(i,k-1)} + \frac{y_3}{V(i+1,k-1)} \ge \frac{1}{V(i,k)}, \tag{4.7}$$

$$\frac{y_1}{V(i-1,k-1)} + \frac{\lambda y_2}{V(i,k-1)} + \frac{y_3}{V(i+1,k-1)} \ge \frac{1}{V(i,k)}, \tag{4.8}$$

$$\frac{y_1}{V(i-1,k-1)} + \frac{y_2}{V(i,k-1)} + \frac{\lambda y_3}{V(i+1,k-1)} \ge \frac{1}{V(i,k)}, \tag{4.9}$$

$$\frac{y_1}{V(i-1,k)} + \frac{y_2}{V(i,k)} + \frac{y_3}{V(i+1,k)} \ge \frac{1}{V(i,k)}. \tag{4.10}$$

4.1.1 Auxiliary Results

First we will formulate an auxiliary Lemma 4.1.1 in which a double sequence $V(i,k)$ is defined and its properties are described. Then in Theorem 4.1.1 will be proved that this sequence gives the value of this game and in Theorem 4.1.2 its limit properties will be described.

Lemma 4.1.1 *Let the double sequence $V(i,k)$ be the solution of the recurrent equation*

$$\frac{1}{V(i,k)} = \max\left\{ \frac{1+\lambda}{V(i-1,k-1)+V(i,k-1)}, \frac{2+\lambda}{V(i-1,k-1)+V(i,k-1)+V(i+1,k-1)} \right\} \tag{4.11}$$

with initial conditions

$$V(i,0) = 1, \quad i = 1,2,\dots, \quad V(1,k) = 1, \quad k = 1,2,\dots. \tag{4.12}$$

Then

$$V(i,k) = (3/(2+\lambda))^k, \quad i = k+1, k+2, \dots, \quad k = 0,1,\dots, \tag{4.13}$$

$$\lambda V(i+1,k) \le V(i,k) \le V(i+1,k), \quad i=1,2,\dots, \quad k=0,1,\dots, \tag{4.14}$$

$$V(i,k) \le V(i,k+1), \quad i=1,2,\dots, \quad k=0,1,\dots, \tag{4.15}$$

$$V(i,k+1) \le (3/(2+\lambda))V(i,k), \quad i=1,2,\dots, \quad k=0,1,\dots. \tag{4.16}$$

In the following table value of $V(i,k)$ for $i \in [1,4], k \in [0,3]$ are given.

	$i=1$	2	3	4
$k=0$	1	1	1	1
1	1	$\frac{3}{2+\lambda}$	$\frac{3}{2+\lambda}$	$\frac{3}{2+\lambda}$
2	1	$\frac{8+\lambda}{(2+\lambda)^2}$	$\left(\frac{3}{2+\lambda}\right)^2$	$\left(\frac{3}{2+\lambda}\right)^2$
3	1	$\frac{12+5\lambda+\lambda^2}{(1+\lambda)(2+\lambda)^2}$	$\frac{26+\lambda}{(2+\lambda)^2}$	$\left(\frac{3}{2+\lambda}\right)^3$

Proof of Lemma 4.1.1 Prove (4.13) by mathematical induction on k. By (4.12), it holds for $k=0$. Assume now that (4.13) holds for a non-negative k. Prove that then it continues to hold for $k+1$. By (4.11) and the assumption, we have

$$\begin{aligned} 1/V(i,k+1) &= \max\{(1+\lambda)/2, (2+\lambda)/3\}\,(3/(2+\lambda))^{-k} \\ &= (3/(2+\lambda))^{-k-1}. \end{aligned}$$

This completes the proof of (4.13).

Pass on to the proof of (4.14) by mathematical induction on k. By (4.12) it holds for $k=0$. Assume now that (4.14) holds for a k. Prove that then it keeps to hold for $k+1$. By (4.11) and the assumption we have

$$\begin{aligned} &\frac{1}{V(i+1,k+1)} \\ &= \max\left\{\frac{1+\lambda}{V(i,k)+V(i+1,k)}, \frac{2+\lambda}{V(i,k)+V(i+1,k)+V(i+2,k)}\right\} \\ &\left\{\begin{matrix}\le 1\\ \ge \lambda\end{matrix}\right\} \max\left\{\frac{1+\lambda}{V(i-1,k)+V(i,k)}, \frac{2+\lambda}{V(i-1,k)+V(i,k)+V(i+1,k)}\right\} \\ &\left\{\begin{matrix}\le 1\\ \ge \lambda\end{matrix}\right\} \frac{1}{V(i,k+1)}. \end{aligned}$$

Hence, $V(i+1,k+1) \ge V(i,k+1) \ge \lambda V(i+1,k+1)$. This completes the proof of (4.14).

Pass on to the proof of (4.15) by mathematical induction on k. By (4.12) and (4.13) it holds for $k = 0$. Assume now that (4.15) holds for a k. Prove that then it keeps to hold for $k + 1$. By (4.11) and the assumption

$$\frac{1}{V(i,k+2)} = \max\left\{\frac{1+\lambda}{V(i-1,k+1)+V(i,k+1)}, \right.$$
$$\left.\frac{2+\lambda}{V(i-1,k+1)+V(i,k+1)+V(i+1,k+1)}\right\}$$
$$\leq \max\left\{\frac{1+\lambda}{V(i-1,k)+V(i,k)}, \right.$$
$$\left.\frac{2+\lambda}{V(i-1,k)+V(i,k)+V(i+1,k)}\right\} = \frac{1}{V(i,k+1)}.$$

Hence, $V(i,k+2) \geq V(i,k+1)$. This completes the proof of (4.15).

Pass on to the proof of (4.16) by mathematical induction on k. By (4.12) and (4.13) it holds for $k = 0$. Assume now that (4.16) holds for a k. Prove that then it keeps to hold for $k + 1$. By (4.11) and the assumption

$$\frac{1}{V(i,k+2)} = \max\left\{\frac{1+\lambda}{V(i-1,k+1)+V(i,k+1)}, \right.$$
$$\left.\frac{2+\lambda}{V(i-1,k+1)+V(i,k+1)+V(i+1,k+1)}\right\}$$
$$\geq \frac{2+\lambda}{3}\max\left\{\frac{1+\lambda}{V(i-1,k)+V(i,k)}, \right.$$
$$\left.\frac{2+\lambda}{V(i-1,k)+V(i,k)+V(i+1,k)}\right\} = \frac{2+\lambda}{3}\frac{1}{V(i,k+1)}.$$

Hence, $V(i,k+2) \leq 3V(i,k+1)/(2+\lambda)$. This completes the proof of (4.16) and Lemma 4.1.1.

Remark 4.1.1 *Let* $\gamma \in \{\leq, >\}$. *Then*

$$\frac{1+\lambda}{V(i-1,k)+V(i,k)} \quad \gamma \quad \frac{2+\lambda}{V(i-1,k)+V(i,k)+V(i+1,k)}$$

if and only if

$$(1+\lambda)V(i+1,k) \quad \gamma \quad V(i-1,k)+V(i,k).$$

4.1.2 The Main Resuls

Theorem 4.1.1 *The value of the game $\Gamma(i,k)$ is given by (4.3), where $V(i,k)$ is given by (4.11), (4.12). The optimal strategies x, y of Guard and Infiltrator on the first step of the game $\Gamma(i,k)$ are given as follows*
(a) if

$$\frac{1}{V(i,k)} = \frac{1+\lambda}{V(i-1,k-1)+V(i,k-1)}, \tag{4.17}$$

then

$$\begin{aligned} x_1 &= \frac{1}{1-\lambda}\frac{V(i,k-1)-\lambda V(i-1,k-1)}{V(i-1,k-1)+V(i,k-1)},\\ x_2 &= \frac{1}{1-\lambda}\frac{V(i-1,k-1)-\lambda V(i,k-1)}{V(i-1,k-1)+V(i,k-1)},\\ x_3 &= x_4 = 0, \end{aligned} \tag{4.18}$$

$$\begin{aligned} y_1 &= \frac{V(i-1,k-1)}{V(i-1,k-1)+V(i,k-1)},\\ y_2 &= \frac{V(i,k-1)}{V(i-1,k-1)+V(i,k-1)},\\ y_3 &= 0, \end{aligned} \tag{4.19}$$

(b) if

$$\frac{1}{V(i,k)} = \frac{2+\lambda}{V(i-1,k-1)+V(i,k-1)+V(i+1,k-1)} \tag{4.20}$$

then

$$\begin{aligned} x_1 &= \frac{1}{1-\lambda}\\ &\times\frac{-(1+\lambda)V(i-1,k-1)+V(i,k-1)+V(i+1,k-1)}{V(i-1,k-1)+V(i+1,k-1)+V(i,k-1)},\\ x_2 &= \frac{1}{1-\lambda}\\ &\times\frac{V(i-1,k-1)-(1+\lambda)V(i,k-1)+V(i+1,k-1)}{V(i-1,k-1)+V(i,k-1)+V(i+1,k-1}, \\ x_3 &= \frac{1}{1-\lambda}\\ &\times\frac{V(i-1,k-1)+V(i,k-1)-(1+\lambda)V(i+1,k-1)}{V(i-1,k-1)+V(i,k-1)+V(i+1,k-1)},\\ x_4 &= 0, \end{aligned} \tag{4.21}$$

$$\begin{aligned} y_1 &= \frac{V(i-1,k-1)}{V(i-1,k-1)+V(i,k-1)+V(i+1,k-1)},\\ y_2 &= \frac{V(i,k-1)}{V(i-1,k-1)+V(i,k-1)+V(i+1,k-1)},\\ y_3 &= \frac{V(i+1,k-1)}{V(i-1,k-1)+V(i,k-1)+V(i+1,k-1)}. \end{aligned} \tag{4.22}$$

Proof (a) Non - negativity of x follows from (4.14). So, the vectors x and y given by (4.18) and (4.19) are probability vectors. The inequalities (4.4), (4.5), (4.7) and (4.8) are satisfied as equalities. By Remark 4.1.1 and (4.17) we have

$$\frac{1}{V(i+1,k-1)} \leq \frac{1+\lambda}{V(i-1,k-1)+V(i,k-1)}.$$

Hence, by (4.18)

$$\frac{x_1+x_2+\lambda x_3}{V(i+1,k-1)} + \frac{x_4}{V(i+1,k)} = \frac{1}{V(i+1,k-1)} \leq \frac{1}{V(i,k)},$$

This implies (4.6). By (4.17) and (4.19) we have

$$\begin{aligned} &\frac{y_1}{V(i-1,k-1)} + \frac{y_2}{V(i,k-1)} + \frac{\lambda y_3}{V(i+1,k-1)} \\ &= \frac{2}{V(i-1,k-1)+V(i,k-1)} \geq \frac{1}{V(i,k)}. \end{aligned}$$

So, (4.9) holds. By (4.16) and (4.19) we have

$$\begin{aligned} &\frac{y_1}{V(i-1,k)} + \frac{y_2}{V(i,k)} + \frac{y_3}{V(i+1,k)} \geq \frac{6/(2+\lambda)}{V(i-1,k-1)+V(i,k-1)} \\ &\geq \frac{1+\lambda}{V(i-1,k-1)+V(i,k-1)} = \frac{1}{V(i,k)}. \end{aligned}$$

Hence, (4.9) also holds. This completes the proof of (a).

(b) Non - negativity of x follows from (4.11), (4.20), Lemma 4.1.1 and Remark 4.1.1. So, the vectors x and y given by (4.21) and (4.22) are probability ones. The inequalities (4.4) - (4.8) are satisfied as equalities. By (4.16) and (4.22) we have

$$\begin{aligned} &\frac{y_1}{V(i-1,k)} + \frac{y_2}{V(i,k)} + \frac{y_3}{V(i+1,k)} \\ &\geq \frac{9/(2+\lambda)}{V(i-1,k-1)+V(i,k-1)+V(i+1,k-1)} \\ &\geq \frac{2+\lambda}{V(i-1,k-1)+V(i,k-1)+V(i+1,k-1)} \\ &= \frac{1}{V(i,k)}. \end{aligned}$$

So, (4.9) holds. This completes the proof of (b) and Theorem 4.1.1.

Theorem 4.1.2 *The following limits hold*
(a) $\lim_{i\to\infty} V(i,k) = (3/(2+\lambda))^k$ *and*

$$\lim_{i\to\infty} x(i,k) = (1/3,1/3,1/3,0), \quad \lim_{i\to\infty} y(i,k) = (1/3,1/3,1/3),$$

(b) $\lim_{k\to\infty} V(i,k) = \lambda^{1-i}$ *and*

$$\lim_{k\to\infty} x(i,k) = (1,0,0,0), \quad \lim_{k\to\infty} y(i,k) = (\frac{\lambda}{1+\lambda}, \frac{1}{1+\lambda}, 0).$$

Proof (a) immediately follows from (4.13), (4.21) and (4.22).

(b) By (4.15) there exists $\lim_{k\to\infty} V(i,k)$. Let

$$V(i) = \lim_{k\to\infty} V(i,k)\,.$$

Then, by (4.11) and (4.12) we have that $V(0) = 1$ and

$$\frac{1}{V(i)} = \max\left\{\frac{1+\lambda}{V(i-1)+V(i)}, \frac{2+\lambda}{V(i-1)+V(i)+V(i+1)}\right\}.$$

Prove that

$$\frac{1}{V(i)} = \frac{1+\lambda}{V(i-1)+V(i)} \quad \text{for} \quad i > 1\,.$$

Assume that there exists a $i_0 \geq 1$ such that

$$\frac{1+\lambda}{V(i_0-1)+V(i_0)} < \frac{2+\lambda}{V(i_0-1)+V(i_0)+V(i_0+1)}$$

Then by the assumption and (4.14) we have

$$\frac{1}{V(i_0+1)} > \frac{1+\lambda}{V(i_0-1)+V(i_0)} \geq \frac{1+\lambda}{V(i_0)+V(i_0+1)}\,.$$

Hence,

$$\frac{1}{V(i_0+1)} = \frac{2+\lambda}{V(i_0)+V(i_0+1)+V(i_0+2)}\,.$$

Repeating this arguments we obtain that

$$\frac{1}{V(i)} = \frac{2+\lambda}{V(i-1)+V(i)+V(i+1)} \quad \text{for} \quad i \geq i_0\,.$$

Hence,

$$(V(i-1)-\lambda V(i)) + (V(i+1)-V(i)) = 0 \quad \text{for} \quad i \geq i_0\,.$$

Then, using (4.14), it implies that

$$V(i-1) = \lambda V(i) \quad \text{and} \quad V(i+1) = V(i) \quad \text{for} \quad i \geq i_0\,.$$

This leads to a contradiction. Hence, $V(i) = V(i-1)/\lambda$ for $i > 1$. So,

$$V(i) = \lambda^{1-i} \quad \text{for} \quad i \geq 1\,.$$

Then (4.18) and (4.19) imply that

$$\lim_{k\to\infty} x(i,k) = (1,0,0,0), \quad \lim_{k\to\infty} y(i,k) = \left(\frac{\lambda}{1+\lambda}, \frac{1}{1+\lambda}, 0\right).$$

This completes the proof of (b) and Theorem 4.1.2.

4.2 A Multi - Stage Customs and Smuggler Game

In this Section we consider a problem of a patrol trying to stop Smuggler who is attempting to ship a cargo of perishable contraband across a strait. Formulate this problem as a zero-sum game of exhaustion in the following way. There are two players: Customs and Smuggler. Smuggler has a small motorboat at his disposal and has to make an attempt to cross a strait to ship contraband on one of n nights. Customs has a speedboat to stop Smuggker. Customs has limited resources and can patrol only during k of n nights. On each night the following variant of the players' behavior are possible:

1. Customs does not assign a patrol and Smuggler ships its cargo of contraband. The game is over.
2. Customs assigns a patrol and Smuggler ships its cargo. Then Smuggler will be caught with probability α and in any case (the capture of Smuggler occurs or not) the game is over.
3. In the cases Smuggler decides not to cross the strait the game is not over but goes on.

There is a restriction on Smuggler's behavior, he must attempt to cross the strait with certainty during one of n nights. The payoff to Customs is 1 if Smuggler is caught and 0 otherwise. We denote by Γ_k^n this game when Smuggler has to attempt to cross the strait during the following n nights and Customs can assign a patrol for k of them. Let $v_k^n = val(\Gamma_k^n)$. Then the game can be described as follows

$$v_k^n = \text{val}\begin{array}{c} \\ \text{patrol} \\ \text{no patrol}\end{array}\begin{array}{c}\begin{array}{cc}\text{go} & \text{don't go}\end{array}\\ \begin{pmatrix} \alpha & v_{k-1}^{n-1} \\ 0 & v_k^{n-1} \end{pmatrix}\end{array}$$

and

$$v_0^n = 0, \quad v_n^n = \alpha,$$

That is, if Customs has no opportunity to patrol the strait then Smuggler freely ships his cargo, and if Customs has no restriction on his resources than Smuggler will be caught with probability α. Then

$$v_k^n = \frac{\alpha k}{n} \quad \text{for} \quad k \in [0, n]$$

and

$$x_1 = \frac{k}{n}, \quad x_2 = 1 - \frac{k}{n},$$
$$y_1 = \frac{1}{n}, \quad y_2 = 1 - \frac{1}{n},$$

where $x_1(x_2)$ is the probability that Customs patrols (does not patrol) during the first night, $y_1(y_2)$ is the probability that Smuggler goes (does go) during the first night.

Consider the two-boat variant of the Customs and Smuggler game, where Customs has two speedboats at its disposal, and it can use its speedboat i during k_i of n nights where $i = 1, 2$. If it patrols with speedboat i during the night when Smuggler chooses to attempt to cross the strait, the capture probability is equal to α_i. If during this night Customs uses both speedboats for patrolling, then the capture probability is $\alpha > \max\{\alpha_1, \alpha_2\}$. We denote by $\Gamma^n_{k_1,k_2}$ this game when Smuggler has to attempt to cross the strait during the following n nights and Customs can assign a patrol with the speedboat i for k_i of them. Let $v^n_{k_1,k_2} = val(\Gamma^n_{k_1,k_2})$. Then the game can be described as follows

if $k_1 k_2 > 0$ then

$$v^n_{k_1,k_2} = \text{val} \begin{matrix} \\ \text{patrol by boat 1} \\ \text{patrol by boat 2} \\ \text{patrol by both boats} \\ \text{no patrol} \end{matrix} \begin{matrix} \text{go} & \text{don't go} \\ \end{matrix} \begin{pmatrix} \alpha_1 & v^{n-1}_{k_1-1,k_2} \\ \alpha_2 & v^{n-1}_{k_1,k_2-1} \\ \alpha & v^{n-1}_{k_1-1,k_2-1} \\ 0 & v^{n-1}_{k_1,k_2} \end{pmatrix},$$

if $k \in [0, n]$, then

$$v^n_{k,0} = v^n_{0,k} = v^n_k .$$

These boundaty conditions are the conditions of the coordination of the one- and two-boat games.

Let x_1, x_2 and x_3 is the probability that Customs patrols by boat 1, boat 2 and both boats, respectively, during the first night, x_4 - is the probability that Customs does not patrol during the first night and $y_1 (y_2)$ is the probability that Smuggler goes (does go) during the first night. Then the probability vectots $x = (x_1, x_2, x_3, x_4), y = (y_1, y_2)$ are the optimal strategies of Customs and Smuggler, respectively, and $v^n_{k_1,k_2}$ is the value of the game $\Gamma^n_{k_1,k_2}$ if and only if the following inequalities hold

$$y_1\alpha_1 + y_2 v^{n-1}_{k_1-1,k_2} \le v^n_{k_1,k_2} , \tag{4.23}$$

$$y_1\alpha_2 + y_2 v^{n-1}_{k_1,k_2-1} \le v^n_{k_1,k_2} , \tag{4.24}$$

$$y_1\alpha + y_2 v^{n-1}_{k_1-1,k_2-1} \le v^n_{k_1,k_2} , \tag{4.25}$$

$$y_2 v^{n-1}_{k_1,k_2} \le v^n_{k_1,k_2} , \tag{4.26}$$

$$v^n_{k_1,k_2} \le x_1\alpha_1 + x_2\alpha_2 + x_3\alpha , \tag{4.27}$$

$$v^n_{k_1,k_2} \le x_1 v^{n-1}_{k_1-1,k_2} + x_2 v^{n-1}_{k_1,k_2-1} + x_3 v^{n-1}_{k_1-1,k_2-1} + x_4 v^{n-1}_{k_1,k_2} . \tag{4.28}$$

4.2.1 The Main Result on Two Boats Game

Theorem 4.2.1 *Let $0 \le k_2 \le k_1 \le n$, then $y_1 = 1/n, y_2 = 1 - 1/n$ and*
(a) if $\alpha \ge \alpha_1 + \alpha_2$, then

$$v^{n}_{k_1,k_2} = \frac{k_2\alpha + (k_1 - k_2)\alpha_1}{n}$$

and

$$x_1 = \frac{k_1 - k_2}{n}, \quad x_2 = 0, \quad x_3 = \frac{k_2}{n}, \quad x_4 = \frac{n - k_1}{n},$$

(b) if $\alpha < \alpha_1 + \alpha_2$ and $k_1 + k_2 < n$, then

$$v^{n}_{k_1,k_2} = \frac{k_1\alpha_1 + k_2\alpha_2}{n}$$

and

$$x_1 = \frac{k_1}{n}, \quad x_2 = \frac{k_2}{n}, \quad x_3 = 0, \quad x_4 = \frac{n - k_1 - k_2}{n},$$

(c) if $\alpha < \alpha_1 + \alpha_2$ and $k_1 + k_2 \ge n$, then

$$v^{n}_{k_1,k_2} = \frac{(k_1 + k_2 - n)\alpha + (n - k_2)\alpha_1 + (n - k_1)\alpha_2}{n}$$

and

$$x_1 = \frac{n - k_2}{n}, \quad x_2 = \frac{n - k_1}{n}, \quad x_3 = \frac{k_1 + k_2 - n}{n}, \quad x_4 = 0.$$

Proof (a) It is easy to see that the inequalities (4.23) and (4.25)-(4.28) are satisfied as equalities. Since $\alpha_1 + \alpha_2 < \alpha$ we have

$$y_1\alpha_2 + y_2 v^{n-1}_{k_1,k_2-1} = \frac{\alpha_1 + \alpha_2 - \alpha}{n} + v^{n}_{k_1,k_2} \le v^{n}_{k_1,k_2}.$$

So, the inequality (4.24) also holds.

(b) The inequalities (4.23), (4.24) and (4.26)-(4.28) are satisfied as equalities. Since $\alpha_1 + \alpha_2 \ge \alpha$ we have

$$y_1\alpha + y_2 v^{n-1}_{k_1-1,k_2-1} = \frac{\alpha - \alpha_1 - \alpha_2}{n} + v^{n}_{k_1,k_2} \le v^{n}_{k_1,k_2}.$$

So, the inequality (4.25) also holds.

(c) The inequalities (4.23)-(4.25), (4.27) and (4.28) are satisfied as equalities. Since $\alpha_1 + \alpha_2 \ge \alpha$ we have

$$y_2 v^{n-1}_{k_1,k_2} = \frac{\alpha - \alpha_1 - \alpha_2}{n} + v^{n}_{k_1,k_2} \le v^{n}_{k_1,k_2}.$$

So, the inequality (4.26) also holds. This completes the proof of Theorem 4.2.1.

4.3 Further Reading

Isaacs [56] investigated a multi-stage infiltration game where Guard has unlimited ammunition. Lee [69],[69] introduced a limit on Guard's supply and solved the game where the payoff to Guard is the number of shots that hit Infiltrator. Nakai [75] solved the infiltration game with unlimited Guard's supply. Sakaguchi [93] suggested to investigate this game with restriction to Guard's supply that made the game more motivated. Baston and Bostock [11] studied the infiltration game under assumptions that Guard has single shot and there is a barrier at point 0 instead of the bunker. So, there is no point where Infiltrator is immune to the Guard's shooting. Garnaev [42] introduced fragmentation shots. Namely, if Guard hits the point where Infiltrator locates, say i, then the probability of hitting Infiltrator is α_1, if Guard hits one of the neighbouring points ($i-1$ or $i+1$) then the probability of hitting Infiltrator is α_2, where $0 < \alpha_2 < \alpha_1 < 1$, and 0 otherwise. Let $\lambda_r = \bar{\alpha}_r, r = 1, 2$ then
if $\lambda_1 + 1 > 2\lambda_2$, then

$$\frac{1}{V(i,k)} = \max\left\{\frac{\lambda_1}{V(i-1,k-1)}, \frac{\lambda_1+\lambda_2}{V(i-1,k-1)+V(i,k-1)},\right.$$

$$\left.\frac{\lambda_2(1-\lambda_1)}{(1-\lambda_2)V(i-1,k-1)+(\lambda_2-\lambda_1)V(i+1,k-1)}\right\},$$

if $\lambda_1 + 1 \le 2\lambda_2$, then

$$\frac{1}{V(i,k)} = \max\left\{\frac{\lambda_1}{V(i-1,k-1)}, \frac{\lambda_1+\lambda_2}{V(i-1,k-1)+V(i,k-1)},\right.$$

$$\frac{\lambda_1+1}{V(i-1,k-1)+V(i+1,k-1)},$$

$$\left.\frac{2\lambda_2^2-\lambda_1^2-\lambda_1}{(\lambda_2-\lambda_1)(V(i-1,k-1)+V(i+1,k-1))+(2\lambda_2-\lambda_1-1)V(i,k-1)}\right\}.$$

Thomas and Nisgav [108] suggested to describe the Customs and Smuggler problem by multi-stage game and solved its one-boat variant. Baston and Bostock [14] solved this game for two-boat Customs and Smuggler game. Garnaev [47] found the solution of its three-boat variant.

Problems

It is interesting to note that in the one-, two- and three-boat Customs and Smuggler games Smuggler has the same optimal strategy: go with probability

$1/n$ and wait with probability $1-1/n$. Is this true for the general p—boat case?

Solve the Customs and Smuggler game for the case where Smuggler has to cross the strait during m of n nights. The the value of the game $v^{n}_{k,m}$ can be found as follows:

$$v^{n}_{k,m} = \text{val}\ \begin{matrix} \text{patrol} \\ \text{no patrol} \end{matrix} \overset{\begin{matrix}\text{go} & \qquad\text{don't go}\end{matrix}}{\begin{pmatrix} \alpha + \bar{\alpha} v^{n-1}_{k-1,m-1} & v^{n-1}_{k-1,m} \\ v^{n-1}_{k,m-1} & v^{n-1}_{k,m} \end{pmatrix}}$$

and

$$v^{n}_{0,m} = 0, \quad v^{n}_{n,m} = \alpha + \bar{\alpha} v^{n}_{n,m-1}, \quad v^{n}_{k,0} = 0\,.$$

Solve the Infiltration game for the case where Infiltrator can move with speed at most r and Guard has k fragmentation shots.

4.4 Inspection Game

Owen [77] considered a game called the Inspection game which has many features in common with the Customs and Smuggler game. In this game there are two players: Inspector and Evador. Evador wants to perform one illegal action during one of the n days. Inspector, wishing to prevent this illegal action, can perform an inspection during one of the n days. The payoff to Inspector is 1 if Evador is caught, 0 if Evador does not attempt to act illegal and -1 if Evador acts illegal uncaught. Evador is definitely caught if he attemps to act illegal when Inspector inspects. Let v^{n} be the value of the game. Then the game can be described as follows

$$v^{n} = \text{val}\ \begin{matrix} \text{inspect} \\ \text{don't inspect} \end{matrix} \overset{\begin{matrix}\text{act} & \text{don't act}\end{matrix}}{\begin{pmatrix} 1 & -1 \\ -1 & v^{n-1} \end{pmatrix}}$$

and $v^{1} = 0$. Then

$$v^{n} = -\frac{n-1}{n+1}\,.$$

Inspector and Evador have the same optimal strategy: to act on the first day with probability $\frac{n^2+1}{n(n+1)}$ and to postpone their actions with probability $\frac{n-1}{n(n+1)}$.

Let Inspector can make k inspection during n days. Denote by v^{n}_{k} the value of this game. Then the game can be described as follows

$$v_k^n = \text{val} \begin{array}{c} \\ \text{inspect} \\ \text{don't inspect} \end{array} \begin{array}{c} \text{act} \quad \text{don't act} \\ \begin{pmatrix} 1 & v_{k-1}^{n-1} \\ -1 & v_k^{n-1} \end{pmatrix} \end{array}$$

and $v_n^n = 0, v_0^n = -1$. Dresher [29], Baston and Bostock [14], Sakaguchi [99], and Ferguson and Melolidakis [32] suggested different proofs of the following Theorem.

Theorem 4.4.1 *The value of the Inspection game is*

$$v_k^n = -\frac{\binom{n-1}{k}}{\sum_{i=0}^{k} \binom{n}{i}}, \tag{4.29}$$

where

$$\binom{n}{k} = \frac{n!}{k!(n-k)!}.$$

The optimal strategy x^* of Inspector is to inspect on the first day with probability $\frac{v_k^{n-1} + v_k^n}{v_{k-1}^{n-1} + v_k^{n-1}}$ and the optimal strategy y^* of Evador is to act illegal on the first day with probability $\frac{v_{k-1}^{n-1} + v_k^{n-1} - v_k^n}{v_{k-1}^{n-1} + v_k^{n-1}}$.

Proof Let

$$w_k^n = (1 - v_k^n)/2,$$

then

$$w_0^n = 1, \quad w_n^n = 1/2 \tag{4.30}$$

and

$$w_k^n = \text{val} \begin{pmatrix} 1 & 0 \\ w_{k-1}^{n-1} & w_k^{n-1} \end{pmatrix}.$$

By induction on n we have

$$1/2 \le w_k^n \le w_{k-1}^n \le 1.$$

So,

$$w_k^n = \frac{w_{k-1}^{n-1}}{1 + w_{k-1}^{n-1} - w_k^{n-1}}. \tag{4.31}$$

Recalling that

$$\binom{n-1}{j-1} + \binom{n-1}{j} = \binom{n}{j} \tag{4.32}$$

we obtain that if (4.29) holds then

$$w_k^n = \frac{\omega_k^{n-1}}{\omega_k^n}, \tag{4.33}$$

where

$$\omega_k^n = \sum_{j=0}^{k} \binom{n}{j}.$$

Note that (4.30) holds for w_k^n given by (4.33). So, it is left to prove that (4.31) also holds for w_k^n given by (4.33). By (4.31) we have that

$$\frac{1}{w_k^{n-1}}\left(\frac{1}{w_k^n} - 1\right) = \frac{1}{w_{k-1}^{n-1}}\left(\frac{1}{w_k^{n-1}} - 1\right).$$

Subsituting (4.33) implies

$$\frac{\omega_k^{n-1}}{\omega_k^{n-2}}\left(\frac{\omega_k^n}{\omega_k^{n-1}} - 1\right) = \frac{\omega_{k-1}^{n-1}}{\omega_{k-1}^{n-2}}\left(\frac{\omega_k^{n-1}}{\omega_k^{n-2}} - 1\right).$$

Hence,

$$\omega_{k-1}^{n-2}(\omega_k^n - \omega_k^{n-1}) = \omega_{k-1}^{n-1}(\omega_k^{n-1} - \omega_k^{n-2}).$$

By (4.32) we have that

$$\omega_k^n - \omega_k^{n-1} = \omega_{k-1}^{n-1},$$
$$\omega_k^{n-1} - \omega_k^{n-2} = \omega_{k-1}^{n-2}.$$

So,

$$\omega_{k-1}^{n-2}\omega_{k-1}^{n-1} = \omega_{k-1}^{n-1}\omega_{k-1}^{n-2}.$$

It implies that (4.31) holds for w_k^n given by (4.33). This completes the proof of Theorem 4.4.1.

It is worth to note that Ruckle [89] suggested a wide range of Inspection games, and Avenhaus, Canty, Kilgour, Stengel and Zamir [7] gave an extended review on Inspection games.

4.5 Game with a Safe Zone, I

In this Section the following Infiltration game on a line is considered. There are two players: Infiltrator and Guard. The game is played on an integer interval $[0, n+1]$ in discrete time $t = 0, 1, ..., T$. At the first instant of time $t = 0$, Infiltrator is located in the safe zone at $x = 0$ and his objective is to reach the sensitive zone at $x = n + 1$ within a time limit T moving with at most unit speed. Guard, having a gun and k shots, where $k \leq T-1$, is located at the sensitive zone during the game and attempts to thwart Infiltrator by shooting him on route to the sensitive zone. The safe and sensitive zones are not available for the Guard's fire. There is no aiming or ballistic error Guard can shoot at any point he desires. Guard's marksmanship is given by the probability $1-\lambda, \lambda \in [0,1)$, of a hit if he shoots at the correct point (i.e. where Infiltrator is). The payoff to Infiltrator is 1 if he reaches the sensitive zone and 0 otherwise. Both players are deemed to have perfect recall of their own actions, and similarly both have complete knowledge of the parameters of the game that they are playing: the payoff and any other parameters which define the game (the time limit on the duration of play for example). Infiltrator never possesses any more information than this - he is deemed to be blind and deaf to the action of his opponent, until such time as he is shot. Guard also has no information on Infiltrator's position.

Without loss of generality we can restrict our attention to only those Infiltrator pure strategies for which the following conditions are satisfied

1. Infiltrator never leaves $x = n+1$ once he has reached it
2. He never goes back to $x = 0$ once he has left it
3. He never goes to a point from which he would have no chance of reaching $x = n+1$ within the time limit.

Definition 4.5.1 *(i) A pure strategy of Infiltrator is a function $I, I : [0,T] \to [0, n+1]$ such that*

1. There are $t_1, t_2 \in [0,T]$ such that $I(t) = 0, t \in [0, t_1], I(t) = n+1, t \in [t_2, T]$ and $I(t) \in [1,n], t \in [t_1+1, t_2-1]$
2. $|I(t) - I(t+1)| \leq 1$, for $t \in [0, T-1]$.

(ii) A pure strategy of Guard is a function $J, J : [0, T-1] \to [1,n] \cup \{\infty\}$ such that $|\{t : J(t) \notin \{\infty\}| = k$. For convenience, we consider that Guard can shoot at ∞ (say, at the sky) without any success.

Let $E(I,J)$ be the payoff to Infiltrator when Infiltrator and Guard use the pure strategies I and J respectively. So, $E(I,J)$ is the probability of successful infiltration. If Infiltrator and Guard employ pure strategies I and J, respectively then the payoff to Infiltrator $E(I,J)$ equals $\lambda^{<I,J>}$ where $<I,J>$ denotes the number of times such that $I(t) = J(t)$ for $t \geq 0$.

Definition 4.5.2 *The i-th Wait-and-Run pure strategy of Infiltrator is denoted by I_i, where*

$$I_i(t) = \begin{cases} 0 & \text{for } t \in [0, i-1], \\ t-i+1 & \text{for } t \in [i, i+n-1], \\ n+1 & \text{for } t \in [i+n, T], \end{cases}$$

where $i \in [1, w], w = T - n$ is the total number of Wait-and-Run strategies. It is easy to see that using the strategy I_i Infiltrator leaves the safe zone at *time $t = i - 1$, then going with unit speed reaches the sensitive zone at time $t = i + n$ and stays there.*

Lemma 4.5.1 *Let I_* be the mixed strategy of Infiltrator saying him to play each of the w Wait-and-Run pure strategies with equal probabilities. Then for each pure strategy of Guard J we have*

$$E(I_*, J) \geq v(k, w, \lambda),$$

where

$$v(k, w, \lambda) = \frac{\lambda^{z+1} s + \lambda^z (w - s)}{w}, \quad s = k - zw, \quad z = \left[\frac{k}{w}\right].$$

First prove the following auxiliary remark.

Remark 4.5.1 *Let a, b and m be non-negative integer such that $a = b + m$ and $\lambda^{a-1} + \lambda^{b+1} \geq \lambda^a + \lambda^b$ then $m = 0$ or 1.*

Proof of Remark 4.5.1 Since $a = b + m$ we have the following sequence of equivalent inequalities

$$\begin{aligned} \lambda^{a-1} + \lambda^{b+1} &\geq \lambda^a + \lambda^b, \\ \lambda^{b+m-1} + \lambda^{b+1} &\geq \lambda^{b+m} + \lambda^b, \\ 0 &\geq (1 - \lambda^{m-1})(1 - \lambda). \end{aligned}$$

Hence, by $\lambda < 1$, we have that $m = 0$ or $m = 1$. This completes the proof of Remark 4.5.1.

We shall say that Infiltrator, following some pure strategy, will be met or shot at by Guard's strategy, if there exists an instant of time when Guard shoots at the position of Infiltrator.

Proof of Lemma 4.5.1 Let Guard and Infiltrator employ strategies J and I_*, respectively. Note that, trajectories of the wait-and-run strategies do not intersect each other in $[1, n]$ and since Infiltrator is avaiable for guard's shots only in this interval $[1, n]$ then only one wait-and-run strategy can be shot at per a shot.

Let $q_i = < I_i, J >$, then $q_i \in [0, n]$,

$$\sum_{i=1}^{w} q_i \leq k$$

and

$$E(I_*, J) = \frac{\sum_{i=1}^{w} \lambda^{q_i}}{w}.$$

It is clear that

$$E(I_*, J) \geq Q ,$$

where

$$Q = \min\left\{\frac{\sum_{i=1}^{w} \lambda^{q_i}}{w} : \sum_{i=1}^{w} q_i \leq k \text{ and } q_i \geq 0 \text{ for } i \in [1, w]\right\}. \qquad (4.34)$$

Prove that

$$Q = v(k, w, \lambda). \qquad (4.35)$$

First prove that if $\{q_i\}$ minimizes (4.34) then

$$\sum_{i=1}^{w} q_i = k. \qquad (4.36)$$

Suppose that this assertion is wrong, then there are $\{q_i\}$, such that the minimum of (4.35) reaches at $\{q_i\}$ such that

$$\sum_{i=1}^{w} q_i < k.$$

Then there exist $\{q_{*i}\}$ such that
(a) $q_{*i} \geq q_i$,
(b) at least one of the inequalities in (a) holds as strong inequality
(c) $\sum_{i=1}^{w} q_{*i} = k$.
Then, since $\lambda < 1$, we have

$$\sum_{i=1}^{w} \lambda^{q_{*i}} < \sum_{i=1}^{w} \lambda^{q_i^j}.$$

This contradiction proves (4.36).

Prove that if $\{q_i\}$ minimizes (4.34), then there is a non-negative integer ξ such that q_i equals either ξ or $\xi + 1$.

Suppose that the assertion is wrong then, there are q_{i*} and $q_{i'}$ such that $q_{i'} - q_{i*} = m \geq 2$. Let

$$\bar{q}_i = \begin{cases} q_i & \text{for } i \notin \{i', i*\}, \\ q_{i*} + 1 & \text{for } i = i*, \\ q_{i'} - 1 & \text{for } i = i'. \end{cases}$$

So, $\bar{q}_i \geq 0$ and

$$\sum_{i=1}^{w} \bar{q}_i = k .$$

Hence, by Remark 4.5.1 and the assumption that $m \geq 2$, we have

$$\sum_{i=1}^{w} \lambda^{\bar{q}_i} < \sum_{i=1}^{w} \lambda^{q_i} .$$

This contradiction proves that there exists a described ξ. Let α of $\{q_i\}$ equal $\xi + 1$, and $w - \alpha$ of them equal ξ, then

$$(\xi + 1)\alpha + \xi(w - \alpha) = k .$$

Hence,

$$Qw = f(\xi) = \alpha\lambda^{\xi+1} + (w - \alpha)\lambda^{\xi} ,$$

where $\alpha = k - w\xi, \xi \in [0, [k/w]]$. It is clear that

$$f(\xi) \geq f([k/w]) \quad \text{for} \quad \xi \in [0, [k/w]] .$$

This completes the proof of (4.35) and Lemma 4.5.1.

Remark 4.5.2 *Let $w \geq k$.*
(a) If $\{1 \leq \alpha_1 < \ldots \alpha_w \leq T - 1\}$ then $\alpha_i \in [i, i + n - 1], i \in [1, w]$.
(b) Let a pure strategy of Guard (say, J) is to shoot at point $I_i(\alpha_i)$ at time $t = \alpha_i$. Then irrespective of Infiltrator's motion he will be met by this Guard's strategy at least once.

Proof (a) follows from the following sequences of inequalities:

$$\alpha_i \geq \alpha_{i-1} + 1 \geq \ldots \geq \alpha_1 + i - 1 \geq (\text{by } \alpha_1 \geq 1) \geq i ,$$

$$\alpha_i \leq \alpha_{i+1} - 1 \leq \ldots \leq \alpha_w + i - w \leq (\text{by } \alpha_w \leq T - 1 \text{ and } w = T - n) \leq i + n - 1 .$$

(b) By (a) J is correctly defined and meets any Wait-and-Run strategy of Infiltrator. Suppose that there is a strategy I of Infiltrator not being met by J. Let Φ be the subset of all pure Infiltrator strategies not being met by J. Suppose that $I^* \in \Phi$ coincide with a Wait-and-Run strategy during the maximal initial time interval (say I_i, where $i \in [1, w]$). Then there is $\tau \in [1, T - 1]$ such that

$$I^*(t) = I_i(t), t \in [1, \tau] \text{ and } \quad I^*(\tau + 1) \neq I_i(\tau + 1)$$

Let $s > i$ be such that

$$I^*(\tau + 1) = I_s(\tau + 1)$$

then $I^\bullet \in \Phi$, where

$$I^\bullet(t) = \begin{cases} I_i(t) & \text{for } t \leq \tau + 1 , \\ I^*(t) & \text{for } t > \tau + 1 \end{cases}$$

and $I^\bullet$ coincides with a Wait-and-Run strategy during time interval $[1, \tau + 1]$. This contradicts the definition of I^* and completes the proof of Remark 4.5.2.

Lemma 4.5.2 *Let $J_i, i \in [1, w]$ be pure Guard strategies given as follows*

$$J_i(t) = \begin{cases} 1 & \text{for } t \in [i, i+s-1] \bmod w\,, \\ I_r(t) & \text{for } t = \alpha_j^r, r \in [1, w],\ j \in [1, k]\,, \end{cases} \tag{4.37}$$

where $[1, n] \backslash \{[i, i+s-1] \bmod w\} = \{\alpha_1^1 < \ldots < \alpha_w^1 < \alpha_1^2 < \ldots < \alpha_w^2 < \ldots < \alpha_w^k\}$ and for any integer a, b and c by $[a, b] \bmod c$ the following integer set $\{x \in [1, c] :$ there is $y \in [a, b]$ such that $x = y \bmod c\}$ is denoted. Then irrespective of Infiltrator's motion

(a) Infiltrator will be met at least k times by each of the Guard's strategies $\{J_i\}$,

(b) Infiltrator will be met at least $k+1$ times by at least s of the Guard's strategies.

Proof (a) follows from the bottom line of (4.37) and Remark 4.5.2.

(b) Let Infiltrator use a pure strategy. Then he has to pass by the point $x = 1$ within time $[1, w]$. Therefore, by the top line of (4.37) Infiltrator will be met at $x = 1$ at least once at least s of the $\{J_i\}$. So, by the bottom line of (4.37) and (a) he will be met at least $k+1$ times by at least s of them. This completes the proof of Lemma 4.5.2.

Lemma 4.5.3 *Let J_* be the mixed strategy of Guard saying him to play each of the w $\{J_i\}$ strategies with equal probabilities. Then for each pure strategy of Infiltrator I we have*

$$E(I, J_*) \leq v(k, w, \lambda)\,.$$

Proof By Lemma 4.5.2 we have that for any pure strategy I Infiltrator can do not better then be met $k+1$ times by s of the $\{J_i\}$ and k times by $w-s$ of them. So,

$$E(I, J_*) \leq \frac{\lambda^{z+1} s + \lambda^z (w-s)}{w} = v(k, w, \lambda)\,.$$

This completes the proof of Lemma 4.5.3.

From Lemmas 4.5.1 and 4.5.3 we have

Theorem 4.5.1 *The value of the game is $v(k, w, \lambda)$. The optimal strategies of Infiltrator and Guard are I_* and J_*, respectively.*

4.6 Game with a Safe Zone, II

In this Section the other version of the Gal problem on a line is considered. There are two players: Infiltrator and Guard. The game is played on an integer interval $[0, n+1]$ in discrete time $t = 0, 1, ..., T$. At time $t = 0$, Infiltrator and

Guard are located at $x = 0$ and $x = 1$, respectively. They can move with speed at most unit within interval $[0, n+1]$ and $[1, n]$, respectively. So, a pure strategy of Infiltrator is defined by Definition 1. A pure strategy of Guard is a function $J, J : [0, T-1] \to [1, n]$ such that $J(0) = 1$ and $|J(t) - J(t+1)| \le 1$ for $t \in [0, T-2]$. The probability of capture is $1 - \lambda$ for each time in which both players occupy the same point (independently of previous history) and zero otherwise. Players have no information on the opponent's position. At $x = 0$ a safe zone is located for Infiltrator where he is not avaiable for Guard. Also, Guard can not make ambush at the sensitive zone located at $x = n+1$. Infiltrator is to reach it within a time limit T. The payoff to Infiltrator is 1 if he reaches the sensitive zone not being caught and zero otherwise.

From Lemma 4.5.1 we have

Lemma 4.6.1 *Let I_* be the mixed strategy of Infiltrator saying him to play each of the w Wait-and-Run pure strategies with equal probabilities. Then for each pure strategy of Guard J we have*

$$E(I_*, J) \ge v(T-1, w, \lambda).$$

Following the proofs of Remark 4.5.2 and Lemma 4.5.2 we have

Lemma 4.6.2 *(a) Let $\delta_i = 0$ or $\delta_i = 1$ for $i \in [1, w]$ be such that $\sum_{i=1}^{w} \delta_i = s$. Also, let $J_{\delta_1,\dots,\delta_w}$ be a pure strategy of Guard such that*

$$J_{\delta_1,\dots,\delta_w}(t) = I_i(t) \quad \text{for} \quad t \in [1 + (i-1)z + \sum_{j=0}^{i-1} \delta_j, iz + \sum_{j=0}^{i} \delta_j], \quad i \in [1, w].$$

Then irrespective of Infiltrator's motion he will be met by this Guard's strategy at least z times.

(b) Let $J_i = J_{\delta_1^i,\dots,\delta_w^i}$, $i \in [1, w]$ where

$$\delta_j^i = \begin{cases} 1 & \text{for } j \in [i, i+s-1] \bmod w, \\ 0 & \text{otherwise}. \end{cases}$$

Then irrespective of Infiltrator's motion

(a) Infiltrator will be met at least k times by each of the the Guard's strategies $\{J_i\}$,

(b) Infiltrator will be met at least $k+1$ times by at least s of the the Guard's strategies $\{J_i\}$.

From Lemma 4.6.2 we have

Lemma 4.6.3 *Let J_* be the mixed strategy of Guard saying him to play each of the w $\{J_i\}$ strategies with equal probabilities. Then for each pure strategy of Infiltrator I we have*

$$E(I, J_*) \leq v(T-1, w, \lambda) .$$

From Lemmas 4.6.1 and 4.6.3 we have

Theorem 4.6.1 *The value of the game is $v(T-1, w, \lambda)$. The optimal strategies of Infiltrator and Guard are I_* and J_*, respectively.*

4.7 Game Without Safe Zone

In this Section we consider the game of previous section without assumptions that

(a) a safe zone is located at $x = 0$,
(b) Infiltrator must reach the sensitive zone within a time limit.

Namely, suppose that the game is played in discrete time $t = 0, 1, \ldots$ on an integer interval $[1, n+1]$. At $t = 0$ both players are located at $x = 1$. Infiltrator and Guard can move with speed at most unit within interval $[1, n+1]$ and $[1, n]$, respectively. Additionally, for simplification we assume that players cannot move back. Infiltrator's objective is to reach a sensitive zone located at $x = n+1$ without being captured by Guard. A necessary condition for capture is that Infiltrator and Guard be at the same point at the same time instant but, even when these necessary conditions are satisfied, capture only occurs with probability $1-\lambda$, where $\lambda \in [0, 1)$. Both players know their mutual locations at time $t = 0$ but neither player receive any subsequent information about his opponent's position unless capture occurs. We call this game The Simplest Infiltration Game. It is natural to consider only those Infiltrator pure strategy using which he reaches $x = n+1$ for finite time. So, a pure strategy of Infiltrator is a function $I : [0, \infty) \to [0, n+1]$ such that $I(0) = 1, I(t+1) \geq I(t) \geq I(t+1) - 1, t \in [0, \infty)$ and $I(t) \leq n$ for $t < t_i$ and $I(t) = n+1$ for $t \geq t_i$ where $t_i \in [1, \infty)$ depends on I. A pure strategy of Guard is a function $J : [0, \infty) \to [1, n]$ such that $J(0) = 1, J(t+1) \geq J(t) \geq J(t+1) - 1, t \in [0, \infty)$. The payoff to Infiltrator is 1 if he reaches the sensitive zone without being captured and 0 otherwise. Denote this game by Γ_n.

Definition 4.7.1 *(a) The k-th Wait-and-Run strategy of Guard denoted by J_k is to remain at $x = 1$ until time $t = k$ then run directly to $x = n$ and remain there for good.*
(b) The k-th Wait-and-Run strategy of Infiltrator denoted by I_k is to remain at $x = 1$ until time $t = k$ then run directly to $x = n+1$ and remain there for good.

Consider an auxiliary game Γ_n^r where both players employ only Wait-and-Run strategies. This game can be analyzed as an iterated game. Denote the value of this game by v_n^r and consider the situation at time $t = 0$.

1. If both players choose to run then the payoff to Infiltrator is λ^n since he will have exactly n locations with Guard before reaching the point $x = n$.
2. If Infiltrator chooses to run and Guard chooses to wait then the payoff to Infiltrator is λ since he will elude Guard only at time $t = 0$.
3. If Infiltrator chooses to wait and Guard chooses to run then the payoff to Infiltrator is λ^2 since he will encounter Guard only at points $x = 1$ and $x = n$.
4. If both players choose to wait then at time $t = 1$ they will be engaged in the same game again except that Guard has had one search opportunity. Thus, the payoff to Infiltrator is λv_n^r.

Then the game matrix for Γ_n^r is

		Guard	
		Run	Wait
Infiltrator	Run	λ^n	λ
	Wait	λ^2	λv_n^r

Denote by p_I and p_G the probabilities that Infiltrator and Guard respectively run at time $t = 0$. Then

$$\begin{aligned} v_n^r &= p_I\lambda^n + (1-p_I)\lambda^2\,, \\ v_n^r &= p_I\lambda + (1-p_I)\lambda v_n^r\,, \\ v_n^r &= p_G\lambda^n + (1-p_G)\lambda\,, \\ v_n^r &= p_G\lambda^2 + (1-p_G)\lambda v_n^r\,. \end{aligned} \tag{4.38}$$

Solving the system of the equation (4.38) we have that
(a) p_I is the greater root of the quadratic equation

$$\left\{\lambda^2(1-\lambda^{n-2})\right\}(1-p_I)^2 - \left\{(1-\lambda^{n-1})(1+\lambda)\right\}(1-p_I) + \left\{1-\lambda^{n-1}\right\} = 0\,,$$

(b) p_G is the greater root of the quadratic equation

$$\left\{\lambda(1-\lambda^{n-1})\right\}(1-p_G)^2 - \left\{(1-\lambda^{n-1})(1+\lambda)\right\}(1-p_G) + \lambda(1-\lambda^{n-2}) = 0\,.$$

Definition 4.7.2 *Let $I_*(J_*)$ be the mixed strategy of Infiltrator (Guard) saying him to use the k-th Wait-and-Run strategy with probability* $p_{Ik} = p_I(1-p_I)^k$ $(p_{Gk} = p_G(1-p_G)^k)$ *for* $k = 0, 1, \ldots$.

Remark 4.7.1 *It easy to see that:*
(a)

$$E(I_i, J_j) = \begin{cases} \lambda^{i+1} & \textit{for } i < j\,, \\ \lambda^{i+n} & \textit{for } i = j\,, \\ \lambda^{j+2} & \textit{for } i > j\,, \end{cases}$$

(b)

$$E(I_i, J_*) = E(I_*, J_j) = v_n^r\,,$$

(c) In their optimal behavior Infiltrator never waits at point $x = n - 1$ and Guard never waits at point $x < n - 1$ for good.

Proof (a) and (c) are obvious.
(b) It is easy to see that (a) implies

$$\begin{aligned} E(I_i, J_*) &= \textstyle\sum_{j=0}^{\infty} E(I_i, J_j) p_G (1 - p_G)^j \\ &= \lambda^n p_G + \lambda \textstyle\sum_{j=1}^{\infty} p_G (1 - p_G)^j = \lambda^n p_G + \lambda(1 - p_G) = v_n^r\,, \end{aligned}$$

and by (a) and definition of p_G for $i \geq 1$ we have

$$\begin{aligned} E(I_{i+1}, J_*) &- E(I_i, J_*) \\ &= \textstyle\sum_{j<i+1} \lambda^{j+2} p_{Gj} + \lambda^{i+1+n} p_{G(i+1)} + \lambda^{i+2} \sum_{j>i+1} p_{Gj} \\ &- \textstyle\sum_{j<i} \lambda^{j+2} p_{Gj} - \lambda^{i+n} p_{Gi} - \lambda^{i+1} \sum_{j>i} p_{Gj} \\ &= \lambda^{i+2} p_{Gi} + \lambda^{i+1+n} p_{G(i+1)} + \lambda^{i+2} (1 - p_G)^{i+1} \\ &- \lambda^{i+n} p_{Gi} - \lambda^{i+1} (1 - p_G)^i \\ &= \lambda^{i+1} (1 - p_G)^{i-1} (\{\lambda(1 - \lambda^{n-1})\}(1 - p_G)^2 \\ &- \{(1 - \lambda^{n-1})(1 + \lambda)\}(1 - p_G) + \lambda(1 - \lambda^{n-2})) = 0\,. \end{aligned}$$

So, $E(I_i, J_*) = v_n^r$. Analogously we can prove that $E(I_*, J_j) = v_n^r$. This completes the proof of Remark 4.7.1.

Theorem 4.7.1 *Let $n = 2$. Then the value of the game Γ_2 is $v_2^r = \lambda^2$. Optimal strategies of Infiltrator and Guard are I_* and J_*, respectively, and $p_I = \lambda/(1 + \lambda)$, $\quad p_G = 1$.*

Proof By Remark 4.7.1(c) for $n = 2$ players can restrict themselves only to the Wait-and-Run strategies. So, Theorem 4.7.1 follows from Remark 4.7.1(b).

Lemma 4.7.1 *Let $n = 3$ and I be a pure strategy of Infiltrator such that*

$$I(t) = \begin{cases} 1 & \textit{for } t \in [0, k]\,, \\ 2 & \textit{for } t \in [k + 1, k + m]\,, \\ 3 & \textit{for } t = k + m + 1\,, \\ 4 & \textit{for } t \geq k + m + 2\,, \end{cases}$$

where $k \geq 0, m \geq 1$. Then

$$E(I, J_*) < E(I_k, J_*) = v_3^r\,.$$

Proof It is clear that
(a) if $k > 0$ then

$$E(I, J_j) = \begin{cases} \lambda^{j+1} & \text{for } j < k\,, \\ \lambda^{k+2} & \text{for } j \in [k, k+m-1]\,, \\ \lambda^{k} & \text{for } i \geq k+m\,, \end{cases}$$

(b) if $k = 0$ then

$$E(I, J_j) = \begin{cases} \lambda^3 & \text{for } j \in [0, m-1]\,, \\ \lambda & \text{for } i \geq m\,. \end{cases}$$

Then, using Remark4.7.1 (a) and (b) we have that
(a) if $k > 0$ then

$$\begin{aligned} & v_3^r - E(I, J_*) \\ & = E(I_k, J_*) - E(I, J_*) \\ & = \textstyle\sum_{j<k} p_{Gj}\lambda^{j+1} + p_{Gk}\lambda^{k+2} + \sum_{j>k} p_{Gj}\lambda^k \\ & - \textstyle\sum_{j<k} p_{Gj}\lambda^{j+1} - \sum_{j\in[k,k+m-1]} p_{Gj}\lambda^{k+2} - \sum_{j\geq k+m} p_{Gj}\lambda^k \\ & = (\lambda^k - \lambda^{k+2}) \textstyle\sum_{j\in[k+1,k+m-1]} p_{Gj} > 0\,, \end{aligned}$$

(b) if $k = 0$ then

$$\begin{aligned} & v_3^r - E(I, J_*) \\ & = E(I_0, J_*) - E(I, J_*) \\ & = p_{G0}\lambda^3 + \textstyle\sum_{j>0} p_{Gj}\lambda \\ & - \textstyle\sum_{j\in[0,m-1]} p_{Gj}\lambda^3 - \sum_{j\geq m} p_{Gj}\lambda \\ & = (\lambda - \lambda^3) \textstyle\sum_{j\in[1,m-1]} p_{Gj} > 0\,. \end{aligned}$$

This completes the proof of Lemma 4.7.1.

Lemma 4.7.2 *Let $n = 3$, and J and J' be pure strategies of Infiltrator such that*

$$J(t) = \begin{cases} 1 & \text{for } t \in [0, k]\,, \\ 2 & \text{for } t \in [k+1, k+m]\,, \\ 3 & \text{for } t \geq k+m+1\,, \end{cases}$$
$$J'(t) = \begin{cases} 1 & \text{for } t \in [0, k]\,, \\ 2 & \text{for } t \in [k+1, k+m-1]\,, \\ 3 & \text{for } t \geq k+m\,, \end{cases}$$

where $k \geq 0, m \geq 1$. Then

$$E(I_*, J) > E(I_*, J')\,.$$

Hence,

$$E(I_*, J) > E(I_*, J_k) = v_3^r\,.$$

Proof It is clear that
(a) if $k > 0$ then

$$E(I_i, J) = \begin{cases} \lambda^i & \text{for } i < k\,, \\ \lambda^{k+2} & \text{for } i = k+m\,, \\ \lambda^{k+1} & \text{otherwise}\,, \end{cases}$$

(b) if $k = 0$ then

$$E(I_i, J) = \begin{cases} \lambda^3 & \text{for } i = m\,, \\ \lambda^2 & \text{otherwise}\,. \end{cases}$$

Hence,
(a) if $k > 0$ then

$$\begin{aligned} &E(I_*, J) - E(I_*, J') \\ &= \textstyle\sum_{i<k} p_{Ii}\lambda^i + p_{I(k+m)}\lambda^{k+2} + \sum_{i>k, i\neq k+m} p_{Ii}\lambda^{k+1} \\ &- \textstyle\sum_{i<k} p_{Ii}\lambda^i - p_{I(k+m-1)}\lambda^{k+2} - \sum_{i\geq k, i\neq k+m-1} p_{Ii}\lambda^{k+1} \\ &= p_{I(k+m-1)}\lambda^{k+1} + p_{I(k+m)}\lambda^{k+2} - p_{I(k+m)}\lambda^{k+1} - p_{I(k+m-1)}\lambda^{k+2} \\ &= p_{I(k+m-1)}\lambda^{k+1}(1-\lambda)p_i > 0, \end{aligned}$$

(a) if $k = 0$ then

$$\begin{aligned} &E(I_*, J) - E(I_*, J') \\ &= p_{Im}\lambda^3 + \textstyle\sum_{i>0, i\neq m} p_{Ii}\lambda^2 \\ &-p_{I(m-1)}\lambda^3 - \textstyle\sum_{i\geq 0, i\neq m-1} p_{Ii}\lambda^2 \\ &= p_{Im}\lambda^3 + p_{I(m-1)}\lambda^2 - p_{I(m-1)}\lambda^3 - p_{Im}\lambda^2 \\ &= p_{I(m-1)}\lambda^2(1-\lambda)p_i > 0\,. \end{aligned}$$

Then repeating these arguments we have that

$$E(I_*, J) > E(I_*, J_k) = (\text{by Remark 4.7.1(b)}) = v_3^r\,.$$

This completes the proof of Lemma 4.7.2.

From Lemmas 4.7.1 and 4.7.2, and Remark 4.7.1(c) we have the following theorem.

Theorem 4.7.2 *Let $n = 3$, then the value of the game Γ_3 is v_3^r. Optimal strategies of Infiltrator and Guard are I_* and J_*, respectively.*

Let $n = 4, \lambda = 0.1$, then $p_i = 0.0834, v_4^r = 0.00917$. Consider a pure strategy of Guard

$$J(t) = \begin{cases} t & \text{for } t \in [1,3]\,, \\ 3 & \text{for } t = 4\,, \\ 4 & \text{for } t > 4\,. \end{cases}$$

Then

$$\begin{aligned} E(I_*, J) &= p_{I0}\lambda^3 + p_{I1}\lambda^3 + \textstyle\sum_{i\geq 2} p_{Ii}\lambda^2 \\ &= (p_i + p_i(1-p_i))\lambda^3 + (1-p_i)^2\lambda^2 = 0.00845 < v_4^r\,. \end{aligned}$$

So, in general for $n \geq 4$ the Wait-and-Run strategy are not optimal.

4.8 Further Reading

Gal [40] proposed a very general zero-sum game involving an Infiltrator and a Guard. Infiltrator enters a region at a given point O at a given time with the objective of reaching a sensitive zone A and Guard, wishing to thwart Infiltrator, tries to maximize the probability of capturing Infiltrator. As a particular "way-in" to the problem, Gal put forward a discrete problem on a line where both players perform motion with at most unit speed. Lalley [67] solved it assuming that: (a) at point O Infiltrator is immune to the Guard's actions and (b) there is a time limit on reaching the sensitive zone by Infiltrator. He introduced the Wait-and-Run strategies of Infiltrator. Further investigation showed that these Wait-and-Run strategies remain being optimal ones for Infiltrator for a whole series of generalization of this game. Namely, Auger [6] solved the Lalley's game without restriction on Guard's speed in the case where the safe zone O and sensitive zone A are connected by n non-intersecting arcs. Alpern [3] investigated this game on arbitrary graphs. Garnaev, Garnaeva and Goutal [52] considered a modification of the Auger's game in which there is a limit on the number of searches that Guard can make. Baston and Garnaev [17] studied the game when Infiltrator can move with speed at most u, where $u \geq 1$. In its original setup without assuming existence of the bunker and the time limit the Gal's game was considered by Ruckle [88]. He introduced an analog of the Wait-and-Run strategies. Pavlovic [78] showed that these Wait-and-Run strategies are not optimal in general.

A zero-sum game of an immobile Hider versus a mobile Searcher on a line being a variant of the Linear Search Problem by Bellman [23] was first formulated by Beck and Newman [22]. In the Beck and Newman game Searcher starts moving from a specified point with unit speed. Hides chooses a hiding point ant stayes there. The payoff to Hider is the time spent until Hider is captured. The capter occures if the locations of the players concide. Beck and Newman solved this game under the assumption that Hider chooses his location by using probability distributions with first absolute moment not exceeding a known constant M.

There are a series of papers on a zero-sum search game with the following plot. Hider chooses an integer point in a segment (discrete or continuous) and hides an object there. Searcher makes a sequence of guesses to detect the object. After each guess he gets some information on detailing its location. Gal [37] solved a discrete search game on $[1, n]$. Searcher tries to locate it by choosing points $x_1, x_2, \ldots$. Alter choosing each point Searcher is told whether his guess equals or greater than the point where the object is. The payoff to Hider is the number of guesses by Searcher till detecting the object. Gal [38] also solved a zero-sum stochastic search game on the segment $[0, 1)$. In this game Hider hides an object at a point, say y, on $[0, 1)$. Searcher tries to locate it by choosing points $x_i, i \in [1, n]$. After choosing each point x_i Searcher asks Hider: Is the object at the point greater than x_i? There is a positive probability of obtaining wrong answer to these questions. After making these n observations, Searcher chooses a set E. The payoff to Searcher is $1/\operatorname{meas}(E)$, where $\operatorname{meas}(E)$ is the Lebesgue measure of E, if $y \in E$ amd 0 otherwise. It is interesting that in both games the optimal strategy of Hider is to hide the object according to uniform distribution. Baston and Bostock [9] suggested another search game on $[0, 1]$. In this game Hider chooses a point, say y in $[0, 1]$ and hides there an object. searcher successively chooses points $x_1, x_2, \ldots$, where at each point x_i he is told whether $y = x_i, y < x_i$ or $y > x_i$ and he may chooses y_{i+1} in the light of this information. The payoff to Hider is $\sum_{i \geq 1} |x_i - y|$. It is interesting to note that uniform distribution is not optimal strategy for Hider. Alpern [2] found pure minimax search strategy giving the best estimation on the cost to find any y.

Problems

Solve the Simplest Infiltration Game for $n \geq 4$.

Solve the Infiltration game if Infiltrator has safe zone at $x = 0$ and $x = k$, where $k \in (2, n - 2)$.

5 Games of Timing

5.1 Non-zero Sum Silent Duel

Consider the following non-zero sum duel. Two players (player 1 and 2), starting at time $t = 0$ at a unit distance before each one's target, walk toward each one's target at a constant unit speed with no opportunity to retreat. They will reach their targets at time $t = 1$. Each player has a gun with one bullet, which may be shot at any time in [0, 1]. Each player selects a time to shoot. The accuracies of shooting are described by the accuracy functions $A_1(x)$ and $A_2(x)$, where $A_i(x)$ is the probability of hitting by player i his target if he shoots at time x. These functions are differentiable and strictly increasing in [0, 1] such that $A_i(0) = 0, A_i(1) = 1$ and $A_i'(x) > 0$ for $x \in (0, 1)$, where $i = 1, 2$. The accuracy functions are fixed and known beforehand to both players. As soon as one of the players hits his target, the contest is over and the first player hitting his target gets payoff 1, and his opponent gets payoff zero. If none of the players hit their targets their payoffs are zero. If both players hit their target at the same time they share their payoffs. Suppose that both players have silent guns, so if player 1 and player 2 shoot at time x and y, respectively, their payoffs are given as follows

$$M_1(x,y) = \begin{cases} A_1(x) & \text{for } x < y\,, \\ P_1(x) & \text{for } x = y\,, \\ (1 - A_2(y))A_1(x) & \text{for } x > y\,, \end{cases}$$

$$M_2(x,y) = \begin{cases} A_2(y) & \text{for } y < x\,, \\ P_2(y) & \text{for } y = x\,, \\ (1 - A_1(x))A_2(y) & \text{for } y > x\,, \end{cases}$$

where $P_i(x) \in [0, A_i(x))$ for $x \in (0, 1]$, $P_i(0) = 0$, for example, if the players share the payoff equally when they both hit the targets, then

$$P_i(x) = A_i(x)(1 - A_j(x)) + A_i(x)A_j(x)/2 = A_i(x)(1 - A_j(x)/2)\,,$$

where $i = 3 - j$ and $j = 1, 2$.

A pure strategy for a player is a time $x \in [0, 1]$ and a mixed strategy is then a cumulative probability distribution function F in [0, 1].

We remind the reader that the support supp F of the distribution function F is

$$\begin{aligned}\operatorname{supp} F = &\left\{x \in (0, 1) : \begin{array}{l}F(x+\epsilon) - F(x-\epsilon) > 0 \text{ for any sufficiently} \\ \text{small positive } \epsilon\end{array}\right\} \\ &\cup \{1, \text{if } F(1-\epsilon) < 1 \text{ for any sufficiently small positive } \epsilon\} \\ &\cup \{0, \text{if } F(0) > 0\}.\end{aligned}$$

For given a mixed strategy F_2 for player 2 and a pure strategy x for player 1, the payoff to player 1 is given by

$$M_1(x, F_2) = \int_{[0,x)} (1 - A_2(y))A_1(x)\,dF_2(y) + \int_{(x,1]} A_1(x)\,dF_2(y) + P_1(x)q_2(x)$$

where

$$q_i(x) = F_i(x) - F_i(x-0) \quad \text{for} \quad i = 1, 2.$$

Then, by right-continuity of distribution functions F_1 and F_2, the payoffs to player 1 and 2 are given by

$$\begin{aligned}M_1(x, F_2) &= A_1(x)\rho_2(x) + P_1(x)q_2(x), \\ M_2(F_1, y) &= A_2(y)\rho_1(y) + P_2(y)q_1(y).\end{aligned} \tag{5.1}$$

where

$$\rho_i(x) = 1 - F_i(x) + \int_{[0,x)} (1 - A_i(y))\,dF_i(y) \quad \text{for} \quad i = 1, 2.$$

Let (η_1, η_2) be the payoff vector corresponding to a pair of mixed strategies (F_1, F_2) of player 1 and 2. Then

$$\begin{aligned}\eta_1 &= \int_0^1 M_1(x, F_2)\,dF_1(x), \\ \eta_2 &= \int_0^1 M_2(F_1, y)\,dF_2(y).\end{aligned} \tag{5.2}$$

Recall that if (F_1, F_2) is a Nash equilibrium with the payoff vector (η_1, η_2) then

$$\begin{aligned}M_1(x, F_2) &\le \eta_1 \quad \text{for any} \quad x \in [0, 1], \\ M_2(F_1, y) &\le \eta_2 \quad \text{for any} \quad y \in [0, 1].\end{aligned} \tag{5.3}$$

Remark 5.1.1 *If (F_1, F_2) is a Nash equilibrium with the payoff vector (η_1, η_2), then $M_1(x, F_2) = \eta_1$ if one of the following conditions hold:*
(a) $x \in \operatorname{supp} F_1$ and $M_1(t, F_2)$ is continuous at $t = x$,
(b) $q_1(x) > 0$.

Proof (a) Assume that the assertion is false. Then $M_1(x, F_2) < \eta_1$. Since $M_1(t, F_2)$ is continuous at $t = x$ there is a sufficient small positive δ such that

$$M_1(t, F_2) < \eta_1 \text{ for } t \in E\,, \tag{5.4}$$

where $E = (x - \delta,\, x + \delta) \cap [0,\, 1]$. Then, using (5.2) implies that

$$\eta_1 = \int_0^1 M_1(t, F_2)\, dF_1(t) = \int_E M_1(t, F_2)\, dF_1(t) + \int_{[0,1]\setminus E} M_1(t, F_2)\, dF_1(t)$$
$$< (\text{by (5.3) and the fact that } x \in \operatorname{supp}(F_1)) < \eta_1\,.$$

This contradiction completes the proof of (a).

(b) Suppose that the assertion is false. So, $M_1(x, F_2) < \eta_1$. Then, by (5.2)

$$\eta_1 = \int_0^1 M_1(t, F_2)\, dF_1(t) = M_1(x, F_2) q_1(x) + \int_{[0,1]\setminus\{x\}} M_1(t, F_2)\, dF_1(t)$$
$$< (\text{since } M_1(x, F_2) < \eta_1) < \eta_1\,.$$

This contradiction completes the proof of (b) and Remark 5.1.1.

Remark 5.1.2 *Since F_i is right-continuous function we have that*

$$\begin{aligned} M_1(x - 0, F_2) &= M_1(x, F_2) + (A_1(x) - P_1(x))q_2(x) \quad \textit{for} \quad x \in (0,\, 1]\,,\\ M_1(x + 0, F_2) &= M_1(x, F_2)\\ &\quad + \Big(A_1(x)(1 - A_2(x)) - P_1(x)\Big)q_2(x) \quad \textit{for} \quad x \in [0,\, 1)\,. \end{aligned}$$

The corresponding relations hold for $M_2(F_1, y)$.

Remark 5.1.3 *If F_2 is constant in an interval $(a,\, b) \subseteq (0,\, 1)$, then $M_1(\cdot, F_2)$ is strictly increasing in $(a,\, b)$.*

Proof Let $x_1,\, x_2 \in (a,\, b)$ with $x_1 < x_2$, then, since F_2 is constant in $(a,\, b)$, we obtain

$$M_1(x_2, F_2) - M_1(x_1, F_2) = (A_1(x_2) - A_1(x_1))\, \rho_2(x_2) > 0\,.$$

This completes the proof of Remark 5.1.3.

Lemma 5.1.1 *Let $(F_1,\, F_2)$ be a Nash equilibrium, then for $j = 3 - i$ and $i = 1,\, 2$*
(a) $q_1(x)q_2(x) = 0$ for $x \in (0,\, 1]$,
(b) $q_i(0) = 0$,
(c) if $(a,\, b) \subseteq (0,\, 1)$ and F_j is constant in $(a,\, b)$, then F_i also is constant in $(a,\, b)$,
(d) if $x \in (0,\, 1)$ such that $q_j(x) = 0$, then $q_i(x) = 0$,

(e) if there is $[a, b) \subseteq (0, 1)$ such that F_i is strictly increasing continuous function $[a, b)$ then

$$F_i'(x) = -\eta_j \left(\frac{1}{A_j(x)}\right)' \frac{1}{A_i(x)} \quad \text{for} \quad x \in (a, b).$$

Proof (a) Suppose that the assertion is false. Then there is $x \in (0, 1]$ such that $q_1(x)q_2(x) > 0$. So, by Remark 5.1.1(b) and (5.3)

$$\eta_1 - M_1(x-0, F_2) = M_1(x, F_2) - M_1(x-0, F_2) = -(A_1(x) - P_1(x))q_2(x) < 0.$$

Hence, there is a sufficiently small positive ω such that $\eta_1 < M_1(x - \omega, F_2)$. This contradicts the fact that (F_1, F_2) is a Nash equilibrium.

(b) First prove that $\eta_i > 0$ for $i = 1, 2$. Let $F_* = 1/2\chi_{[1/3,1]} + 1/2\chi_{[2/3,1]}$, then

$$\begin{aligned} M_1(F_*, y) &= \frac{1}{2}\begin{cases} A_1(1/3) & \text{if } y > 1/3, \\ P_1(1/3) & \text{if } y = 1/3, \\ (1 - A_2(y))A_1(1/3) & \text{if } y < 1/3 \end{cases} \\ &+ \frac{1}{2}\begin{cases} A_1(2/3) & \text{if } y > 2/3, \\ P_1(2/3) & \text{if } y = 2/3, \\ (1 - A_2(y))A_1(2/3) & \text{if } y < 2/3 \end{cases} \\ &= \frac{1}{2}\begin{cases} A_1(1/3) + A_1(2/3) & \text{if } y > 2/3, \\ A_1(1/3) + P_1(2/3) & \text{if } y = 2/3, \\ A_1(1/3) + (1 - A_2(y))A_1(2/3) & \text{if } y \in (1/3, 2/3), \\ P_1(1/3) + (1 - A_2(y))A_1(2/3) & \text{if } y = 1/3, \\ (1 - A_2(y))(A_1(1/3) + A_1(2/3)) & \text{if } y < 1/3. \end{cases} \end{aligned}$$

Hence, there is a positive $\eta > 0$ such that $M_1(F_*, y) > \eta$ for any $y \in [0, 1]$. Thus,

$$M_1(F_*, F_2) = \int_0^1 M_1(F_*, y)\, dF_2(y) \geq \eta.$$

So,

$$\eta_1 = \int_0^1 M_1(y, F_2)\, dF_1(y) = \int_0^1 M_1(F_1, y)\, dF_2(y) \geq M_1(F_*, F_2) \geq \eta > 0.$$

Now pass on to the proof that $q_i(0) = 0$ for $i = 1, 2$. Suppose that the assertion is false, say for $i = 2$. Then $\eta_1 = M_1(0, F_2) = 0$. But $\eta_1 > 0$ and the result follows.

(c) Say $i = 1$, then, by Remark 5.1.3, $M_1(x, F_2)$ is strictly increasing in (a, b). Hence $(a, b) \cap \operatorname{supp} F_1 = \emptyset$. This completes the proof of (c).

(d) Suppose that the assertion is false. Say $i = 2$, then there is $x \in (0,\ 1)$ such that $q_1(x) = 0$ and $q_2(x) > 0$. Then, by Remark 5.1.2

$$M_1(x+0, F_2) - M_1(x-0, F_2) = -A_1(x)A_2(x)q_2(x)\,.$$

Hence, there is a sufficiently small positive ω such that

$$M_1(z_2, F_2) < M_1(z_1, F_2) \quad \text{for any} \quad z_1 \in (x-\omega,\ x),\ \ z_2 \in (x, x+\omega)\,.$$

Hence, by right-continuity, F_1 is constant in $[x,\ x+\omega)$ and so is F_2 by (c). By Remark 5.1.1(b) $M_2(F_1, x) = \eta_2$ and by Remark 5.1.3 $M_2(F_1, x)$ is strictly increasing in $[x,\ x+\omega)$ and we have contradiction.

(e) Say $i = 2$. Then, by Remark 5.1.1(a) $M_1(x, F_2) = \eta_1$ for $x \in (a,\ b)$. Thus,

$$1 - F_2(x) + \int_{[0,x)} (1 - A_2(y))\, dF_2(y) = \eta_1 / A_1(x)\,.$$

So, since F_2 is strictly increasing continuous function $[a,\ b)$, it also is differentiable in $(a,\ b)$. Then

$$F_2'(x) = -\eta_1 \left(\frac{1}{A_1(x)}\right)' \frac{1}{A_2(x)} \quad \text{for} \quad x \in (a,\ b)\,.$$

This completes the proof of (e) and Lemma 5.1.1.

Lemma 5.1.2 *Let $(F_1,\ F_2)$ be a Nash equilibrium, then F_1 and F_2 are continuous in* $[0,\ 1)$. *Furthermore $M_1(x, F_2) = \eta_1$ for $x \in \operatorname{supp} F_1$.*

Proof The result immediately follows from Lemma 5.1.1 and Remark 5.1.1(a).

Lemma 5.1.3 *Let $(F_1,\ F_2)$ be a Nash equilibrium, then $F_i(x) < 1$ for $x \in [0,\ 1)$ and $i = 1,\ 2$.*

Proof Suppose there is $x' \in \operatorname{supp}(F_i)$ such that $F_i(x') = 1$ and $F_i(x) < 1$ for $x < x'$. Say $i = 1$. Then, since $x' \in \operatorname{supp}(F_1)$ by Lemma 5.1.2 we have $M_1(x', F_2) = \eta_1$. By Remark 5.1.3 and Lemma 5.1.1(c) $M_1(x, F_2)$ is strictly increasing in $(x',\ 1)$. This contradiction completes the proof of Lemma 5.1.3.

Lemma 5.1.4 *Let $(F_1,\ F_2)$ be a Nash equilibrium and $0 < a < b < 1$ such that $a, b \in \operatorname{supp} F_i$ then $F_j(a) \neq F_j(b)$ for $j = 3 - i$ and $i = 1,\ 2$.*

Proof Suppose that there is a subinterval $(a,\ b) \subset (0,\ 1)$ such that $a, b \in \operatorname{supp}(F_i)$ and $F_j(a) = F_j(b)$. Say $j = 2$, then by (5.1), Lemma 5.1.2 and Remark 5.1.1(b) we have

$$\eta_1 = M_1(a, F_2) = A_1(a)\rho_2(a) < A_1(b)\rho_2(a) = A_1(b)\rho_2(b) = M_1(b, F_2) = \eta_1\,.$$

This contradiction completes the proof of Lemma 5.1.4.

Lemma 5.1.5 *Let (F_1, F_2) be a Nash equilibrium with the payoff vector (η_1, η_2), then there is $a \in (0, 1)$ such that*

$$F_i(x) \text{ is } \begin{cases} 0 & \text{for } x \in [0, a)\,, \\ \text{strictly increasing and continuous} & \text{in } [a, 1)\,, \\ 1 & \text{for } x = 1\,, \end{cases} \tag{5.5}$$
$$\eta_i = A_i(a)\,,$$

where $i = 1, 2$. Also,

$$M_1(x, F_2) = \begin{cases} A_1(x) & \text{for } x \in [0, a)\,, \\ A_1(a) & \text{for } x \in [a, 1)\,, \\ A_1(a) + (P_1(1) - A_1(1))q_2(1) & \text{for } x = 1\,. \end{cases} \tag{5.6}$$

Similar expression holds for $M_2(F_1, x)$.

Proof Let (F_1, F_2) be a Nash equilibrium. Then, by Lemmas from 5.1.1 to 5.1.4 there exists $a \in [0, 1)$, such that $F_i(x) = 0$ for $x < a$ and $F_i(x) > 0$ for $x > a$ where $i = 1, 2$. Also, by Lemmas 5.1.3 and 5.1.4, F_1, F_2 are strictly increasing continuous function $[a, 1)$. So, (F_1, F_2) are to be given by (5.5). Then, (5.1) and Lemma 5.1.2 imply that $\eta_i = A_i(a)$ and $M_1(x, F_2)$ are to be given by (5.6). This completes the proof of Lemma 5.1.5.

Remark 5.1.4 *Let $s \in (0, 1)$ and*

$$\phi_i(a, s) = -A_j(a) \int_a^s \left(\frac{1}{A_j(y)}\right)' \frac{1}{A_i(y)}\, dy \quad \text{for} \quad j = 3 - i,\ i = 1, 2\,.$$

Then, for $i = 1, 2$
(a) $\phi_i(a, s)$ is a strictly decreasing function on $a \in (0, s]$,
(b) the equation $\phi_i(a, s) = 1$ has the unique root $a = a_i(s) \in (0, s]$.

Proof (a) It is clear that for $a \in (0, s)$ we have that

$$\begin{aligned} \frac{\partial}{\partial a}\varphi_i(a, s) &= -A_j'(a) \int_a^s \left(\frac{1}{A_j(y)}\right)' \frac{1}{A_i(y)}\, dy - \frac{A_j'(a)}{A_j(a)A_i(a)} \\ &= -A_j'(a) \left(\frac{1}{A_j(s)A_i(s)} + \int_a^s \frac{A_i'(y)}{A_j(y)A_i(y)^2}\, dy\right) \\ &< (\text{since } A_1'(x) > 0 \text{ and } A_2'(x) > 0 \text{ for } x \in (0, 1)) < 0. \end{aligned}$$

Hence, $\varphi_i(a, s)$ is strictly decreasing for $a \in (0, s]$.

(b) Note that

$$\varphi_i(s, s) = 0$$

and

$$\begin{aligned}
\varphi_i(a,s) &> (\text{since } A_i'(x) > 0 \text{ for } x \in (0,1)) \\
&\geq \frac{A_j(a)}{A_i(s)} \int_a^s \frac{A_j'(y)}{A_j(y)^2}\,dy \\
&= \frac{A_j(a)}{A_i(s)} \left(\frac{1}{A_j(a)} - \frac{1}{A_j(s)} \right) \to \frac{1}{A_i(s)} > 1 \text{ for } a \to 0 .
\end{aligned}$$

Then (a) implies that the equation $\varphi_i(\cdot, s) = 1$ has unique root in $[0,\, s)$. This completes the proof of Remark 5.1.4.

Theorem 5.1.1 *There is unique Nash equilibrium* $(F_1,\, F_2)$ *in the silent non-zero sum duel with the payoff vector* $(\eta_1,\, \eta_2)$*. Also,*

$$\begin{aligned}
F_i(x) &= \begin{cases} 0 & \text{for } x \in [0,\, a_*)\,, \\ \varphi_i(a_*, x) & \text{for } x \in [a_*,\, 1)\,, \\ 1 & \text{for } x = 1\,, \end{cases} \\
\eta_i &= A_i(a_*)\,,
\end{aligned} \tag{5.7}$$

where $i = 1,\, 2$ *and* $a_* = \max\{a_1(1),\, a_2(1)\}$.

Proof Let $(F_1,\, F_2)$ be a Nash equilibrium with the payoff vector $(\eta_1,\, \eta_2)$. By Lemma 5.1.5 they are to be given by (5.5). Then, using Lemma 5.1.1(e) we obtain that $F_i(x) = \varphi_i(a, x)$ for $x \in [a,\, 1)$, where $a \in \{a_1(1),\, a_2(1)\}$. Suppose that $a = \min\{a_1(1),\, a_2(1)\} < a_*$. Without loss of generality we can assume that $a = a_2(1)$. Then

$$F_1(1) = \varphi_1(a_2(1), 1) > (\text{by Remark 5.1.4(a)}) > \varphi_1(a_1(1), 1) = 1\,.$$

This contradiction proves that a has to be equal to a_*.

Now prove that $(F_1,\, F_2)$ given by (5.7) where $a = a_*$ is Nash equilibrium with the payoff vector $(\eta_1,\, \eta_2)$ given by (5.7). Without loss of generality we can assume that, $a = a_2(1)$, then $q_2(1) = 0$, $q_1(1) \geq 0$. Therefore,

$$\begin{aligned}
\eta_1 &= \int_0^1 M_1(y, F_2)\, dF_1(y) = \int_{[a,1)} M_1(y, F_2)\, dF_1(y) + A_1(a) q_1(1) = A_1(a), \\
\eta_2 &= \int_0^1 M_2(F_1, x)\, dF_2(x) = \int_{[a,1)} M_2(F_1, x)\, dF_2(x) = A_2(a)\,.
\end{aligned}$$

So, (5.2) holds. By (5.6), we have that

$$M_1(x, F_2) \leq A_1(a) = \eta_1 \text{ and } M_2(F_1, x) \leq A_2(a) = \eta_2 \text{ for } x \in [0,\, 1]\,.$$

So, (5.3) also holds. It implies that $(F_1,\, F_2)$ is a Nash equilibrium with $a = a_*$. This completes the proof of Theorem 5.1.1.

Let $A_1(x) = A_2(x) = x$ for $x \in [0,\, 1]$ then $(F,\, F)$ is unique Nash equilibrium with payoff $(a,\, a)$ given as follows

$$F(x) = \begin{cases} 0 & \text{if } x \in [0, a)\,, \\ \frac{a}{2}\left(\frac{1}{a^2} - \frac{1}{x^2}\right) & \text{if } x \in [a, 1]\,, \end{cases}$$

where $a = \sqrt{2} - 1$.

5.2 Non-zero Sum Noisy Duel

Consider a variant of the duel from previous Section where both players have noisy guns. So, player becomes informed of the shooting of his opponent. Then, if the opponent failed to hit the target and the player has not shot, he will wait to shoot till time $t = 1$ when he can hit his target with certainty. So, if player 1 and player 2 schedule to shoot at time x and y, respectively, their payoffs are given as follows

$$M_1(x,y) = \begin{cases} A_1(x) & \text{for } x < y\,, \\ P_1(x) & \text{for } x = y\,, \\ 1 - A_2(y) & \text{for } x > y\,, \end{cases}$$
$$M_2(x,y) = \begin{cases} A_2(y) & \text{for } y < x\,, \\ P_2(y) & \text{for } y = x\,, \\ 1 - A_1(x) & \text{for } y > x\,. \end{cases}$$

For given distribution functions F_1 and F_2 and $x \in [0, 1]$, the payoffs to players are given by

$$M_1(x, F_2) = A_1(x)(1 - F_2(x)) + \int_{[0,x)} (1 - A_2(y))\, dF_2(y) + P_1(x)q_2(x)\,,$$
$$M_2(F_1, x) = A_2(x)(1 - F_1(x)) + \int_{[0,x)} (1 - A_1(y))\, dF_1(y) + P_2(x)q_1(x)\,.$$

Recall that $(F_{1\delta}, F_{2\delta})$ is a δ Nash equilibrium with the payoff vector (η_1, η_2), where $\delta > 0$, if the following relations hold

$$\int_0^1 M_1(x, F_{2\delta})\, dF_{1\delta}(x) \geq \eta_1 - \delta\,,$$
$$\int_0^1 M_2(F_{1\delta}, y)\, dF_{2\delta}(y) \geq \eta_2 - \delta$$

and

$$M_1(x, F_{2\delta}) \leq \eta_1 + \delta \quad \text{for any} \quad x \in [0, 1]\,,$$
$$M_2(F_{1\delta}, y) \leq \eta_2 + \delta \quad \text{for any} \quad y \in [0, 1]\,.$$

Since A_i, $i = 1, 2$ are continuos and strictly increasing in $[0, 1]$ such that

$A_1(0) + A_2(0) = 0$ and $A_1(1) + A_2(1) = 2$ there is unique root x^* of the equation $L(x) = 0$, where

$$L(x) = A_1(x) + A_2(x) - 1\,.$$

Theorem 5.2.1 *(F_δ, F_δ) is $\epsilon(\delta)$ Nash equilibrium in noisy duel with the payoff vector (η_1, η_2), where $\epsilon(\delta) \to 0$ for $\delta \to 0$ and*

$$F_\delta(x) = \begin{cases} 0 & \text{for } x \in [0, x^*), \\ (x - x^*)/\delta & \text{for } x \in [x^*, x^* + \delta), \\ 1 & \text{for } x \in [x^* + \delta, 1], \end{cases}$$

$$\eta_i = A_i(x^*) \quad \text{for} \quad i = 1, 2.$$

Proof It is clear that for sufficient small positive δ we have that

$$M_1(x, F_\delta) = \begin{cases} A_1(x) & \text{for } x \in [0, x^*), \\ \frac{1}{\delta}\{A_1(x)(\delta + x_* - x) + \\ + \int_{x_*}^{x} (1 - A_2(y))\, dy\} & \text{for } x \in [x^*, x^* + \delta), \\ \frac{1}{\delta}\int_{x^*}^{x^*+\delta} (1 - A_2(y))\, dy & \text{for } x \in [x^* + \delta, 1]. \end{cases}$$

Then, definition of x^* implies that

$$M_1(x, F_\delta) \begin{cases} = A_1(x) & \text{for } x \in [0, x^*), \\ \geq A_1(x^*) + L(x^* + \delta) & \text{for } x \in [x^*, x^* + \delta), \\ \geq A_1(x^*) + (A_2(x^*) - A_2(x^* + \delta)) & \text{for } x \in [x^* + \delta, 1] \end{cases}$$

and

$$M_1(x, F_\delta) \begin{cases} \leq A_1(x^*) + A_1(x^* + \delta) - (1 - \delta)A_1(x^*) & \text{for } x \in [x^*, x^* + \delta), \\ \leq A_1(x^*) & \text{for } x \in [x^* + \delta, 1]. \end{cases}$$

Hence,

$$M_1(x, F_\delta) \leq A_1(x^*) + \epsilon(\delta) \quad \text{for} \quad x \in [0, 1]$$

and

$$\int_0^1 M_1(x, F_\delta)\, dF_\delta(x) \geq A_1(x^*) - \epsilon(\delta),$$

where

$$\epsilon(\delta) = \max\{A_2(x^* + \delta) - A_2(x^*),\ A_1(x^* + \delta) - (1 - \delta)A_1(x^*),\ -L(x^* + \delta)\},$$

and $\epsilon(\delta) \to 0$ for $\delta \to 0$. Similar relations hold for $M_2(F_{1\delta}, y)$. So, (F_δ, F_δ) is $\epsilon(\delta)$ Nash equilibrium with payoff vector $(A_1(x^*), A_2(x^*))$. This completes the proof of Theorem 5.2.1.

Let $A_1(x) = A_2(x) = x$ for $x \in [0, 1]$ then (F_δ, F_δ) is $\epsilon(\delta)$ Nash equilibrium with payoff $(1/2, 1/2)$ where

$$F_\delta(x) = \begin{cases} 0 & \text{for } x \in [0, 1/2), \\ (x - 1/2)/\delta & \text{for } x \in [1/2, 1/2 + \delta), \\ 1 & \text{for } x \in [1/2 + \delta, 1]. \end{cases}$$

5.3 Non-zero Sum Silent-Noisy Duel

Consider a variant of the duel from the first section of this chapter where one player (say, 1) has silent gun and the other one (say, 2) has noisy gun. Then if players 1 and 2 schedule to shoot at time x and y, respectively, their payoffs are given by

$$M_1(x,y) = \begin{cases} A_1(x) & \text{for } x < y\,, \\ P_1(x) & \text{for } x = y\,, \\ 1 - A_2(y) & \text{for } x > y\,, \end{cases}$$
$$M_2(x,y) = \begin{cases} A_2(y) & \text{for } y < x\,, \\ P_2(y) & \text{for } y = x\,, \\ A_2(y)(1 - A_1(x)) & \text{for } y > x\,. \end{cases}$$

Theorem 5.3.1 *(F_1, F_2) is Nash equilibrium in noisy-silent duel with the payoff vector (η_1, η_2) where*
(a) if $x^* \geq a_1(1)$ then

$$F_1(x) = \begin{cases} 0 & \textit{for } x \in [0, a)\,, \\ \varphi_1(a, x) & \textit{for } x \in [a, 1)\,, \\ 1 & \textit{for } x = 1\,, \end{cases}$$
$$F_2(x) = \chi_{[a,1]}(x)$$

and

$$\eta_1 = 1 - A_2(a), \quad \eta_2 = A_2(a)\,,$$

where $a \in [a_1(1), x^*]$,
(b) if $x^* < a_1(1)$ then

$$F_1(x) = \begin{cases} 0 & \textit{for } x \in [0, a_1(1))\,, \\ \varphi_1(a_1(1), x) & \textit{for } x \in [a_1(1), 1]\,, \end{cases}$$
$$F_2(x) = \begin{cases} 0 & \textit{for } x \in [0, a_1(1)]\,, \\ \psi(x, a_1(1)) & \textit{for } x \in [a_1(1), 1)\,, \\ 1 & \textit{for } x = 1 \end{cases}$$

and

$$\eta_1 = A_1(a_1(1)), \quad \eta_2 = A_1(a_1(1))\,,$$

where

$$\psi(x,a) = 1 - \exp\left(\int_a^x A_1'(y)/L(y)\,dy\right).$$

Proof (a) It is clear that

$$M_1(x, F_2) = \begin{cases} A_1(x) & \text{for } x \in [0, a)\,, \\ P_1(x) & \text{for } x = a\,, \\ 1 - A_2(a) & \text{for } x \in (a, 1]\,, \end{cases}$$
$$M_2(F_1, x) = \begin{cases} A_2(x) & \text{for } x \in [0, a)\,, \\ A_2(a) & \text{for } x \in [a, 1)\,, \\ A_2(a) + (P_2(1) - A_2(1))q_1(1) & \text{for } x = 1\,. \end{cases}$$

Hence,

$$\eta_1 = \int_0^1 M_1(x, F_2)\, dF_1(x) = 1 - A_2(a),$$
$$\eta_2 = \int_0^1 M_2(F_1, y)\, dF_2(y) = A_2(a)$$

and

$$M_1(x, F_2) \le 1 - A_2(a) \quad \text{for any} \quad x \in [0, 1],$$
$$M_2(F_1, y) \le A_2(a) \quad \text{for any} \quad y \in [0, 1].$$

Therefore, (F_1, F_2) is Nash equilibrium in noisy-silent duel with the payoff vector (η_1, η_2).

(b) It is clear that

$$M_1(x, F_2) = \begin{cases} A_1(x) & \text{for } x \in [0, a_1(1)), \\ A_1(a_1(1)) & \text{for } x \in [a_1(1), 1), \\ A_1(a_1(1)) + (P_1(1) - A_1(1))q_2(1) & \text{for } x = 1, \end{cases}$$
$$M_2(F_1, x) = \begin{cases} A_2(x) & \text{for } x \in [0, a_1(1)), \\ A_2(a_1(1)) & \text{for } x \in [a_1(1), 1]. \end{cases}$$

Hence,

$$\eta_1 = \int_0^1 M_1(x, F_2)\, dF_1(x) = A_1(a_1(1)),$$
$$\eta_2 = \int_0^1 M_2(F_1, y)\, dF_2(y) = A_2(a_1(1))$$

and

$$M_1(x, F_2) \le A_1(a_1(1)) \quad \text{for any} \quad x \in [0, 1],$$
$$M_2(F_1, y) \le A_2(a_1(1)) \quad \text{for any} \quad y \in [0, 1].$$

This completes the proof of Theorem 5.3.1.

Let $A_1(x) = A_2(x) = x$ for $x \in [0, 1]$ then (F_1, F_2) is Nash equilibrium with payoff $(1 - a, a)$ where

$$F_1(x) = \begin{cases} 0 & \text{for } x \in [0, a), \\ \frac{a}{2}\left(\frac{1}{a^2} - \frac{1}{x^2}\right), & \text{for } x \in [a, 1) \\ 1 & \text{for } x = 1, \end{cases}$$
$$F_2(x) = \chi_{[a,1]}(x),$$

where $a \in [\sqrt{2} - 1, 1/2]$.

In zero-sum duels two contestants are to hit each other. Consider, for example, the following classical plot of duel. Two duelists, player 1 and player 2, starting at time t=0 (at a distance 2 unit apart), walk toward each other at a constant unit speed with no opportunity to retreat. They reach each other at time $t = 1$. Each player has a gun with a shot. As soon as one of the players hits his opponent, the contest is over and the payoff of winner (loser) is 1 (-1), respectively. If either none of the players hit his opponent or both players hit each other at the same time or time limit of the game is over their payoffs are zero. Then, for example, in silent duel, if player 1 and player 2 shoot at time x and y, respectively, the payoff to player 1 is given as follows

$$M(x,y) = \begin{cases} A_1(x) - (1 - A_1(x))A_2(y) & \text{for } x < y\,, \\ A_1(x) - A_2(x) & \text{for } x = y\,, \\ (1 - A_2(y))A_1(x) - A_2(y) & \text{for } x > y\,. \end{cases}$$

Methods of solution of zero - sum duels are covered in Karlin [57]. Sakaguchi [92] initiated investigation on non - zero - sum variants of the classical duels.

5.4 A Duel with Random Termination Time

Teraoka [104], [105] and Sakaguchi [94] suggested a generalization of the duel game for the case where time of termination of the game is not fixed but random. Namely, each player can shoot at any time in $[0,T]$, where T is random value in $[0,1]$ with cumulative probability distribution $H(t)$, assumed to be known to both players. Let $K_i(x)$ be the probability that the contest is not yet terminated at time x and player i hits his target when he shoots at time x. So, $K_i(x) = A_i(x)(1 - H(x-0))$. Then in silent game, if player 1 and player 2 shoot at time x and y, respectively, their payoffs are given as follows:

$$\begin{aligned} M_1(x,y) &= \begin{cases} K_1(x) & \text{for } x < y\,, \\ P_1(x) & \text{for } x = y\,, \\ (1 - A_2(y))K_1(x) & \text{for } x > y\,, \end{cases} \\ M_2(x,y) &= \begin{cases} K_2(y) & \text{for } y < x\,, \\ P_2(y) & \text{for } y = x\,, \\ (1 - A_1(x))K_2(y) & \text{for } y > x\,, \end{cases} \end{aligned} \tag{5.8}$$

where Teraoka [104], [105] assumed that $P_i(x) = K_i(x)$ and Sakaguchi [94] supposed that $P_i(x) < K_i(x)$ for $x \in [0,1]$, $i = 1, 2$. The Sakaguchi's assumption seems more involved since shooting by both players at the same time is quite unlikely event. So, payoffs in this event should not effect on the players' behaviour. For example, if players share their payoffs in the case when they hit their target simultaneously, we have that $P_i(x) = K_i(x)(1 - A_j(x)/2)$. Also, Teraoka and Sakaguchi assumed that K_i is differentiable in $[0,1]$, $K_i(0) = K_i(1) = 0$ and there is $m_i \in [0,1]$ such that $K_i'(x) > 0$ for $x \in (0, m_i)$ and $K_i'(x) < 0$ for $x \in (m_i, 1)$, where $i = 1, 2$. For example, if $A_i(x) = x^{a_i}$, where $a_i \geq 1$, and $H(x) = x$ we have that $K_i(x) = x^{a_i}(1-x)$ and $m_i = a_i/(1+a_i)$.

Following Baston and Garnaev [16] solve the Sakaguchi silent game with payoffs given by (5.8) and prove uniqueness of Nash equilibrium. Now pass on to consideration of this game in details.

For distribution functions F_1 and F_1 and a pure strategy x, the payoffs to players are given by

$$\begin{aligned} M_1(x, F_2) &= K_1(x)\rho_2(x) + P_1(x)q_2(x)\,, \\ M_2(F_1, y) &= K_2(y)\rho_1(y) + P_2(y)q_1(y)\,. \end{aligned} \tag{5.9}$$

Recall that for $i = 1,\ 2$

$$\rho_i(x) = 1 - F_j(x) + \int_{[0,x)} (1 - A_j(y))\, dF_j(y)$$

and it is the same as in the non-zero sum silent duel.

Remark 5.4.1 *Since F_j is right-continuous function we have*

$$\begin{aligned} M_1(x-0, F_2) &= M_1(x, F_2) + (K_1(x) - P_1(x))q_2(x) \quad \textit{for} \quad x \in (0,\ 1]\,, \\ M_1(x+0, F_2) &= M_1(x, F_2) \\ &\quad + \Big(K_1(x)(1 - A_2(x)) - P_1(x)\Big)q_2(x) \quad \textit{for} \quad x \in [0,\ 1)\,. \end{aligned}$$

The corresponding relations hold for $M_2(F_1, y)$.

Similar to the silent duel we can prove Remarks from 5.4.2 to 5.4.4, and Lemmas 5.4.1 and 5.4.2 where some properties of Nash equilibrium are given.

Remark 5.4.2 *(i) If F_2 is continuous in an interval $(a,\ b) \subseteq (m_1,\ 1]$ then $M_1(\cdot, F_2)$ is a strictly decreasing function in $(a,\ b)$.*

(ii) If $x > m_1$ and F_2 is continuous at x then there is an arbitrarily small positive δ such that

$$M_1(x, F_2) < M_1(x - \delta, F_2)\,.$$

(iii) If F_2 is constant in a nondegenerate interval I then $M_1(\cdot, F_2)$ is a strictly increasing in I if $I \subseteq [0,\ m_1]$ and strictly decreasing in I if $I \subseteq [m_1,\ 1]$.

Remark 5.4.3 *Let $(F_1,\ F_2)$ be a Nash equilibrium. Then, $\eta_i > 0$ for $i = 1,\ 2$.*

Remark 5.4.4 *Let $s \in (0,\ m_j)$ and*

$$\phi_i(a, s) = -K_j(a) \int_a^s \left(\frac{1}{K_j(y)}\right)' \frac{1}{A_i(y)}\, dy\,.$$

Then
(a) $\phi_i(a, s)$ is a strictly decreasing function on $a \in (0,\ s]$,
(b) the equation $\phi_i(a, s) = 1$ has the unique root $a = a_i(s) \in (0,\ s]$.

Lemma 5.4.1 *Let (F_1, F_2) be a Nash equilibrium. Then*
(a) if $x \in (0, 1)$ then $q_1(x)q_2(x) = 0$,
(b) $q_i(0) = q_i(1) = 0$,
(c) if $(a, b) \subset (0, 1)$ such that $m_i \notin (a, b)$ and F_j is constant in (a, b) then F_i also is constant in (a, b),
(d) if $x \in (0, 1)\backslash\{m_i\}$ such that $q_j(x) = 0$ then $q_i(x) = 0$,
(e) if there is $[a, b) \subset [0, m)$, where

$$m = \min\{m_1, m_2\},$$

such that F_i is strictly increasing and continuous in $[a, b)$ and $F_i(a) = 0$ then

$$F_i'(x) = \frac{\eta_j}{A_i(x)}\left(\frac{1}{K_j(x)}\right)'.$$

Proposition 5.4.1 *If (F_1, F_2) is a Nash equilibrium, then F_1 and F_2 are continuous in $[0, 1]\backslash\{m_1, m_2\}$. Furthermore,*

$$M_1(x, F_2) = \eta_1 \quad \textit{for} \quad x \in \operatorname{supp} F_1\backslash\{m_1, m_2\}.$$

Lemma 5.4.2 *If (F_1, F_2) is a Nash equilibrium, then $F_i(x) < 1$ for $x < m$ and $i = 1, 2$.*

Lemma 5.4.3 *If (F_1, F_2) is a Nash equilibrium, then $F_i(x) = 1$ for $x \geq m_i$ and $i = 1, 2$.*

Proof Without loss of generality, say $m_2 \leq m_1$. By Proposition 5.4.1 and right continuity, F_1 and F_2 are continuous in $[m_1, 1]$ and $[m_2, m_1)$. By Remark 5.4.2(i), $M_1(x, F_2)$ and $M_2(F_1, x)$ are strictly decreasing in $(m_1, 1)$ and then, by Lemma 5.4.1(b), $F_i(x) = 1$ for $x \in [m_1, 1]$. By the corresponding result to Remark 5.4.2(i), $M_2(F_1, x)$ is strictly decreasing in (m_2, m_1), so F_2 is constant in (m_2, m_1) and then so is F_1 by Lemma 5.4.2(c).

Let $F_i(x) = c_i$ in $[m_2, m_1)$. Thus, the result is established if $m_1 = m_2$ or $c_2 = 1$.

Let $m_1 > m_2$ and $c_2 < 1$. Then, by Lemma 5.4.1(a), $c_1 = 1$. Since $q_1(m_1) = 0, q_2(m_1) = 0$ by Lemma 5.4.1(d), and we have a contradiction to $c_2 < 1$. The result now follows.

Lemma 5.4.4 *If (F_1, F_2) is a Nash equilibrium, then*

$$a = \inf\{x : F_1(x) > 0 \quad \textit{or} \quad F_2(x) > 0\} \leq m.$$

Furthermore, if $a < m$ then for $i = 1, 2$, $\eta_i = K_i(a)$ and F_i is strictly increasing in $[a, m]$.

Proof Let (F_1, F_2) be a Nash equilibrium and

$$a = \inf\{x : F_1(x) > 0 \quad \text{or} \quad F_2(x) > 0\}.$$

Then, $a \leq m$, for, if $a > m$, at least one of $M_1(m, F_2) > \eta_1$ and $M_2(F_1, m) > \eta_2$ must hold.

Now, assume that $a < m$. Then, by Lemma 5.4.1(c),

$$\inf\{x : F_1(x) > 0\} = \inf\{x : F_2(x) > 0\}.$$

Since $a \in \operatorname{supp} F_1 \cap \operatorname{supp} F_2$, by Proposition 5.4.1,

$$\eta_1 = M_1(a, F_2) = K_1(a) \quad \text{and} \quad \eta_2 = M_2(F_1, a) = K_2(a).$$

Suppose that there is a nondegenerate interval $(b, c) \subseteq (a, m)$ such that one of the F_i is constant on (b, c), then, so is the other by Lemma 5.4.1(c). We may clearly assume that (b, c) is a maximal such interval in (a, m). Then, by Lemma 5.4.1(c) $b \in \operatorname{supp} F_1 \cap \operatorname{supp} F_2$. In particular, by Proposition 5.4.1, $M_1(b, F_2) = \eta_1$. However Remark 5.4.2(iii) now leads to a contradiction since it implies that $M_1(\cdot, F_2)$ is strictly increasing in $[b, c]$.

5.4.1 Uniqueness of Nash Equilibria

The function ξ introduced in the next lemma will provide us with information about the sign of $M_1(m_1, F_2) - M_1(a, F_2)$ for an important class of distribution functions.

Lemma 5.4.5 *(a) Let*

$$\xi(a) = K_1(m_1)\left(\frac{K_1(a)}{K_1(m_2)} - A_2(m_2)(1 - \varphi_2(a, m_2))\right) - K_1(a).$$

Then, ξ is a strictly decreasing function in $[0, m_2]$ and the equation $\xi(a) = 0$ has a root $a = \alpha_*$ in $[0, m_2]$ if and if the following inequality holds:

$$(1 - A_2(m_2))K_1(m_1) \leq K_1(m_2). \tag{5.10}$$

If a root in $[0, m_2]$ exists, it is unique.

(b) $\varphi_2(\alpha_*, m_2) \leq 1$.

(c) If $m_1 = m_2$ then $\alpha_* = a_2(m_2)$.

Proof (a) Note that $\xi(a)$ is strictly decreasing function in $[0, m_2]$, because

$$\begin{aligned}
\xi'(a) &= K_1'(a)\Bigg(\frac{K_1(m_1)}{K_1(m_2)} - 1 - K_1(m_1)A_2(m_2) \\
&\quad \times \int_a^{m_2} \left(\frac{1}{K_1(y)}\right)' \frac{dy}{A_2(y)} - \frac{K_1(m_1)A_2(m_2)}{K_1(a)A_2(a)}\Bigg) \\
&= -K_1'(a)\left(1 + K_1(m_1)A_2(m_2)\int_a^{m_2} \frac{A_2'(y)}{(K_1(a)A_2(a))^2}\,dy\right) < 0,
\end{aligned}$$

where we have used the integration by parts formula. The result follows because

$$\xi(a_2(m_2)) = K_1(a_1(m_2))\left(\frac{K_1(m_1)}{K_1(m_2)} - 1\right) > 0,$$
$$\xi(m_2) = (1 - A_2(m_2))K_1(m_1) - K_1(m_2).$$

This completes the proof of (a)

(b) Suppose that $\varphi_2(\alpha_*, m_2) > 1$. Then,

$$\xi(\alpha_*) > K_1(\alpha_*)\left(\frac{K_1(m_1)}{K_1(m_2)} - 1\right) \geq 0,$$

giving a contradiction and the proof of (b) is complete.

(c) In this case we have

$$\xi(\alpha_*) = -K_1(m_1)A_2(m_2)(1 - \varphi_2(\alpha_*, m_2)),$$

so $\xi(\alpha_*) = 0$ gives $\varphi_2(\alpha_*, m_2) = 1$. This completes the proof of Lemma 5.4.5.

Theorem 5.4.1 *Let $m_1 = m_2$, then there is unique Nash equilibrium (F_1, F_2) with payoff vector (η_1, η_2). Also, for $i = 1, 2$*

$$F_i(x) = \begin{cases} 0 & \text{if } x \in [0, a), \\ \varphi_i(a, x) & \text{if } x \in [a, m), \\ 1 & \text{if } x \in [m, 1], \end{cases} \tag{5.11}$$
$$\eta_i = K_i(a),$$

where $a = a_* = \max\{a_1(m), a_2(m)\}$.

Proof Let (F_1, F_2) be a Nash equilibrium. Then, by Lemmas 5.4.3 and 5.4.4 and Proposition 5.4.1, there is an $a \in [0, m)$ such that $F_i(x) = 0$ for $x < a$, $F_i(x)$ is strictly increasing and continuous for $x \in [a, m)$ and $F_i(x) = 1$ for $x \geq m$ where $i = 1, 2$. So, by Lemma 5.4.1(e), $F_i(x) = \varphi_i(a, x)$ for $x \in [a, m)$. So, (F_1, F_2) are given by (5.11). By Lemma 5.4.1(a), $q_1(m)q_2(m) = 0$, so $a \in \{a_1(m), a_2(m)\}$. Let $a = \min\{a_1(m), a_2(m)\} < a_*$. Without loss of generality, say $a = a_2(m)$. Then, by Remark 5.4.4(a),

$$F_1(m) = \varphi_1(a_2(m), m) > \varphi_1(a_1(m), m) = 1.$$

So, $a = a_*$. Prove that if $a = a_*$ then (F_1, F_2) given by (5.11) is Nash equilibrium. Note that

$$M_1(x, F_2) = \begin{cases} K_1(x) & \text{if } x \in [0, a)\,, \\ K_1(a) & \text{if } x \in [a, m)\,, \\ K_1(a) + (P_1(m) - K_1(m))q_2(m) & \text{if } x = m\,, \\ K_1(x)\left(\dfrac{K_1(a)}{K_1(m)} - A_2(m_1)q_2(m)\right) & \text{if } x \in (m, 1]\,, \end{cases}$$

$$M_2(F_1, x) = \begin{cases} K_2(x) & \text{if } x \in [0, a)\,, \\ K_2(a) & \text{if } x \in [a, m)\,, \\ K_2(a) + (P_2(m) - K_2(m))q_1(m) & \text{if } x = m\,, \\ K_2(x)\left(\dfrac{K_2(a)}{K_2(m)} - A_1(m_2)q_1(m)\right) & \text{if } x \in (m, 1]\,. \end{cases}$$

Thus, since both functions K_1 and K_2 get their maximum at $x = m$,

$$M_1(x, F_2) \le K_1(a) = \eta_1 \text{ and } M_2(F_1, x) \le K_2(a) = \eta_2 \text{ for } x \in [0, 1]\,.$$

Say $a = a_2(m)$, then $q_2(m) = 0, q_1(m) \ge 0$. So,

$$\begin{aligned} \eta_1 &= \int_0^1 M_1(y, F_2)\, dF_1(y) \\ &= \int_{[a,m)} M_1(y, F_2)\, dF_1(y) + K_1(a)q_1(m) = K_1(a)\,, \\ \eta_2 &= \int_0^1 M_2(F_1, x)\, dF_2(x) = \int_{[a,m)} M_2(F_1, x)\, dF_2(x) = K_2(a)\,. \end{aligned}$$

So, $(F_1,\ F_2)$ is Nash equilibrium for $a = a_*$. Now the result follows.

Now without loss of generality we can consider that $m_2 < m_1$.

Lemma 5.4.6 *Let $(F_1,\ F_2)$ be a Nash equilibrium then either*

$$F_i(x) = \chi_{[m_i,1]}(x) \quad \textit{for} \quad i = 1,\ 2$$

or there is $a \in [0,\ m_2)$ such that

$$\begin{aligned} F_1(x) &= \begin{cases} 0 & \textit{if } x \in [0, a)\,, \\ \varphi_1(a, x) & \textit{if } x \in [a, m_2)\,, \\ \varphi_1(a, m_2) & \textit{if } x \in [m_2, m_1)\,, \\ 1 & \textit{if } x \in [m_1, 1], \end{cases} \\ F_2(x) &= \begin{cases} 0 & \textit{if } x \in [0, a)\,, \\ \varphi_2(a, x) & \textit{if } x \in [a, m_2)\,, \\ 1 & \textit{if } x \in [m_2, 1]\,. \end{cases} \end{aligned} \tag{5.12}$$

In the last case

$$M_2(F_1,x)=\begin{cases} K_2(x) & \text{if } x\in[0,a)\,,\\ K_2(a) & \text{if } x\in[a,m_2)\,,\\ \dfrac{K_2(x)K_2(a)}{K_2(m_2)} & \text{if } x\in[m_2,m_1)\,,\\ \dfrac{K_2(m_1)K_2(a)}{K_2(m_2)} & \\ +(P_2(m_2)-K_2(m_2))q_1(m_2) & \text{if } x=m_1\,,\\ K_2(x)\left(\dfrac{K_2(a)}{K_2(m_2)}\right. & \\ \left.-A_1(m_1)q_1(m_1)\right) & \text{if } x\in(m_1,1] \end{cases} \tag{5.13}$$

and

$$M_1(x,F_2)=\begin{cases} K_1(x) & \text{if } x\in[0,a)\,,\\ K_1(a) & \text{if } x\in[a,m_2)\,,\\ K_1(a)+(P_1(m_2)-K_1(m_2))q_2(m_2) & \text{if } x=m_2\,,\\ K_1(x)\left(\dfrac{K_1(a)}{K_1(m_2)}-A_2(m_2)q_2(m_2)\right) & \text{if } x\in(m_2,1]\,. \end{cases} \tag{5.14}$$

Proof Let (F_1, F_2) be a Nash equilibrium. Then, by Lemmas from 5.4.1 to 5.4.4, there is an $a \in [0, m_2]$ such that $F_1(x) = F_2(x) = 0$ for $x \in [0, a]$, $F_1(x) = 1$ for $x \in [m_1, 1]$ and $F_2(x) = 1$ for $x \in [m_2, 1]$. Also,

(i) If $a = m_2$ then by Lemma 5.4.1, $F_1(x) = 0$ for $x \in [m_2, m_1)$ so $F_i(x) = \chi_{[m_i,1]}(x)$ for $i = 1, 2$ and the result follows.

(ii) If $a < m_2$ then by Lemma 5.4.1, $F_i(x) = \varphi_i(a, x)$ for $x \in [a, m_2)$ and $i = 1, 2$. Also, if $q_2(m_2) = 0$, then by Lemma 5.4.1(d), $q_1(m_2) = 0$. Thus, F_i and M_i for $i = 1, 2$ are given by (5.12)–(5.14).

Theorem 5.4.2 *There is a pure Nash equilibrium if and only if*

$$(1 - A_2(m_2))K_1(m_1) \geq K_1(m_2)\,. \tag{5.15}$$

Also, if (5.15) holds then Nash equilibrium is unique and it is (m_1, m_2) with payoff vector $((1 - A_2(m_2))K_1(m_1),\ K_2(m_2))$.

Proof By Lemma 5.4.6, if (F_1, F_2) is a pure Nash equilibrium, then it is (m_1, m_2). It is clear that

$$M_1(m_1,m_2)=(1-A_2(m_2))K_1(m_1)\,,$$
$$M_1(x,m_2)=\begin{cases} K_1(x) & \text{if } x<m_2\,,\\ P_1(x) & \text{if } x=m_2\,,\\ (1-A_2(m_2))K_1(x) & \text{if } x>m_2\,,\end{cases}$$
$$M_2(m_1,m_2)=K_2(m_2)\,,$$
$$M_2(m_1,y)=\begin{cases} K_2(y) & \text{if } y<m_1\,,\\ P_2(y) & \text{if } y=m_1\,,\\ (1-A_1(m_1))K_2(y) & \text{if } y>m_1\,.\end{cases}$$

Then, it follows that (m_1, m_2) is a Nash equilibrium if and only if (5.15) holds.

Let (5.15) hold and there is a Nash equilibrium (F_1, F_2) distinguished from (m_1, m_2). Then it is given by (5.12) for $a \in [0, m_2)$. Thus, by (5.14) and Lemma 5.4.5,

$$M_1(m_1, F_2) = \xi(a) + K_1(a) > K_1(a) = \eta_1 .$$

Now the result follows.

Theorem 5.4.3 *Let (5.15) do not hold, then there is unique Nash equilibrium (F_1, F_2) with the payoff vector (η_1, η_2). Also, F_1 and F_2 given by (5.12) where $a = \max\{a_1(m_2), \alpha_*\}$ and $\eta_i = K_i(a)$ for $i = 1, 2$.*

Proof Let (F_1, F_2) be a Nash equilibrium. Then, by Lemma 5.4.6, there is $a \in [0, m_2)$ such that F_1 and F_2 are given by (5.12).
Let $a < a_1(m_2)$. Then, by Remark 5.4.4,

$$F_1(m_2) = \varphi_1(a, m_2) > \varphi_1(a_1(m_2), m_2) = 1 .$$

The contradiction implies that $a \geq a_1(m_2)$.
Prove that

$$\text{if } a > a_1(m_2) \text{ then } a = \alpha_* . \tag{5.16}$$

By Remark 5.4.4,

$$F_1(m_2) = \varphi_1(a, m_2) < \varphi_1(a_1(m_2), m_2) = 1 .$$

So, $m_1 \in \operatorname{supp} F_1$ and F_2 is continuous at $x = m_1$. Then, by Lemma 5.4.5,

$$\eta_1 = K_1(a) = M_1(m_1, F_2) = \xi(a) + K_1(a) .$$

So, $\xi(a) = 0$. Thus, $a = \alpha_*$ and (5.16) follows.
Prove that

$$\text{if } a = a_1(m_2) \text{ then } a \geq \alpha_* . \tag{5.17}$$

Since, in this case F_1 and F_2 are continuous at $x = m_1$,

$$M_1(m_1, F_2) = \xi(a) + K_1(a) \leq \eta_1 = K_1(a) .$$

So, by Lemma 5.4.5 $a \geq \alpha_*$, and (5.17) follows.

(5.16) and (5.17) implies that $a = \max\{a_1(m_2), \alpha_*\}$. Prove that then (F_1, F_2) is Nash equilibrium. By (5.13) and (5.14), since $K_i(x)$ gets maximum at $x = m_i$ for $i = 1, 2$

$$
\begin{aligned}
&M_2(F_1, x) \le K_2(a)\,,\\
&M_1(x, F_2) = \begin{cases} K_1(x) & \text{if } x \in [0, a)\,,\\ K_1(a) & \text{if } x \in [a, m_2)\,,\\ K_1(a) + (P_1(m_2) - K_1(m_2))q_2(m_2) & \text{if } x = m_2\,,\\ \le K_1(a) + \xi(a) & \text{if } x \in (m_2, 1] \end{cases}\\
&\qquad\qquad \le K_1(a)\,.
\end{aligned}
$$

Also,

$$
\begin{aligned}
\eta_2 &= \int_0^1 M_2(F_1, x)\, dF_2(x)\\
&= \int_{[a,m_2)} M_2(F_1, x)\, dF_2(x) + M_2(F_1, m_2) q_2(m_2) = K_2(a)
\end{aligned}
$$

and

$$
\eta_1 = \int_0^1 M_1(y, F_2)\, dF_1(y) = \int_{[a,m_2)} M_1(y, F_2)\, dF_1(y) + M_1(m_1, F_2) q_1(m_1)\,.
$$

Note that if $a = \alpha_*$, then

$$
M_1(m_1, F_2) = \xi(\alpha_*) + K_1(\alpha_*) = K_1(\alpha_*) = K_1(a)\,,
$$

and if $a = a_1(m_2)$, then

$$
q_1(m_1) = 0\,.
$$

Thus, $\eta_1 = K_1(a)$ and the result follows.

Theorems from 5.4.1 to 5.4.3 implies the following result.

Theorem 5.4.4 *In the silent non-zero sum game with random termination time there is unique Nash equilibrium.*

Let $A_1(x) = A_2(x) = H(x) = x$ for $x \in [0, 1]$. So, $K_1(x) = K_2(x) = x(1-x)$ then $m_1 = m_2 = m$ and (F, F) is Nash equilibrium with payoff (a, a) where

$$
F(x) = \begin{cases} 0 & \text{for } x \in [0, a)\,,\\ a(1-a)\Big(\ln\Big(\frac{a}{1-a}\frac{1-x}{x}\Big) + \frac{1}{1-a} & \\ +\frac{1}{2a^2} - \frac{1}{1-x} - \frac{1}{2x^2}\Big) & \text{for } x \in [a, 1/2)\,,\\ 1 & \text{for } x \in [1/2, 1]\,, \end{cases}
$$

where a is the unique root in $(0, 1/2)$ of the equation

$$
\frac{1}{a} - \frac{1}{2a^2} + 4 = \ln\left(\frac{a}{1-a}\right).
$$

Suppose that $A_1(x) = A_2(x) = x$ and $P_i(x) = K_i(x)$ for $x \in [0, 1]$ and $i = 1, 2$. Let $x_* \in [1/4, 1/2]$, then

$$M_1(x, x_*) = \begin{cases} x(1-x) & \text{if } x \in [0, x_*], \\ (1-x_*)x(1-x) & \text{if } x \in (x_*, 1], \end{cases}$$
$$\leq \max\{(1-x_*)x_*, (1-x_*)/4\} = (1-x_*)x_* = M_1(x_*, x_*).$$

Similar relations hold for M_2. So, (x_*, x_*) for $x \in [1/4, 1/2]$ are Nash equilibria. Thus, the assumption that $P_i(x) < K_i(x)$ for $x \in [0, 1], i = 1, 2$ is essential for uniqueness of Nash equilibrium.

5.5 A Noisy Duel with Random Termination Time

In the Sakaguchi noisy game if player 1 and player 2 schedule to shoot at time x and y, respectively, their payoffs are given as follows:

$$M_1(x, y) = \begin{cases} K_1(x) & \text{for } x < y, \\ P_1(x) & \text{for } x = y, \\ (1 - A_2(y))K_1(\sigma_1(y)) & \text{for } x > y, \end{cases}$$
$$M_2(x, y) = \begin{cases} K_2(y) & \text{for } y < x, \\ P_2(y) & \text{for } y = x, \\ (1 - A_1(x))K_2(\sigma_2(x)) & \text{for } y > x, \end{cases} \tag{5.18}$$

where for $i = 1, 2$

$$\sigma_i(x) = \begin{cases} m_i & \text{for } x \leq m_i, \\ x & \text{for } x > m_i. \end{cases}$$

Let x_j^* be the unique root in $[0, m_j]$ of the equation $L_j(x) = 0$, where

$$L_j(x) = (1 - A_i(x))K_j(m_j) - K_j(x) \quad \text{for} \quad j = 3 - i, \, i = 1, 2.$$

Theorem 5.5.1 *There is a pure Nash equilibrium in the Sakaguchi noisy duel if and only if*

$$m_1 \leq x_2^* \quad or \quad m_2 \leq x_1^*. \tag{5.19}$$

(a) If $m_1 \leq x_2^*$, then (x_*, y_*) is a pure Nash equilibrium with payoff vector (η_1, η_2) if and only if $(x_*, y_*) = (m_1, y)$, where $y \in (m_1, 1]$ and $\eta_1 = K_1(m_1)$, $\eta_2 = (1 - A_1(m_1))K_2(m_2)$.

(b) If $m_2 \leq x_1^*$, then (x_*, y_*) is a pure Nash equilibrium with payoff vector (η_1, η_2) if and only if $(x_*, y_*) = (x, m_2)$, where $x \in (m_2, 1]$ and $\eta_1 = (1 - A_2(m_2))K_1(m_1)$, $\eta_2 = K_2(m_2)$.

Note that if (5.19) does not hold, then either $x_2^* < x_1^* < m_2$, or $x_1^* < x_2^* < m_2$, or $x_1^* = x_2^* < m_2$.

Without loss of generality we can consider that $m_2 < m_1$.

Theorem 5.5.2 *Let (5.19) do not hold. (F_1, F_2) is a Nash equilibrium with the payoff vector (η_1, η_2) in the Sakaguchi noisy duel where*
(a) if $x_2^* < x_1^* < m_2$, then

$$F_1(x) = \begin{cases} 0 & \text{for } x \in [0, a), \\ \psi_1(x,a) & \text{for } x \in [a, m_2), \\ 1 & \text{for } x \in [m_2, 1], \end{cases}$$
$$F_2(x) = \chi_{[a,1]}(x)$$

and

$$\eta_1 = (1 - A_2(a))K_1(m_1),$$
$$\eta_2 = K_2(a),$$

where $a \in (x_2^*, x_1^*]$ and

$$\psi_i(x,a) = 1 - \exp\left(\int_a^x K_j'(y)/L_j(y)\,dy\right) \quad \text{for} \quad i = 3 - j \quad \text{and} \quad j = 1, 2.$$

(b) if $x_1^* < x_2^* < m_2$, then

$$F_1(x) = \chi_{[a,1]}(x),$$
$$F_2(x) = \begin{cases} 0 & \text{for } x \in [0, a), \\ \psi_2(x,a) & \text{for } x \in [a, m_1), \\ 1 & \text{for } x \in [m_1, 1] \end{cases}$$

and

$$\eta_1 = K_1(a),$$
$$\eta_2 = (1 - A_1(a))K_2(m_2),$$

where $a \in (x_1^*, x_2^*]$.

Theorem 5.5.3 *Let (5.19) do not hold and $x_1^* = x_2^* < m_2$, then (F_δ, F_δ) is $\epsilon(\delta)$ Nash equilibrium with the payoff vector (η_1, η_2), where $\epsilon(\delta) \to 0$ for $\delta \to 0$ and*

$$F_\delta(x) = \begin{cases} 0 & \text{for } x \in [0, x_1^*), \\ (x - x_1^*)/\delta & \text{for } x \in [x_1^*, x_1^* + \delta), \\ 1 & \text{for } x \in [x_1^* + \delta, 1], \end{cases}$$
$$\eta_i = K_i(x_1^*) \quad \text{for} \quad i = 1, 2.$$

5.6 A Silent-Noisy Duel with Random Termination Time

In the Sakaguchi silent-noisy game, where, say, player 1 has silent gun and player 2 has noisy one, if player 1 and player 2 schedule to shoot at time x and y, respectively, their payoffs are given as follows:

$$M_1(x,y)=\begin{cases}K_1(x) & \text{for } x<y\,,\\ P_1(x) & \text{for } x=y\,,\\ (1-A_2(y))K_1(\sigma_1(y)) & \text{for } x>y\,,\end{cases}$$
$$M_2(x,y)=\begin{cases}K_2(y) & \text{for } y<x\,,\\ P_2(y) & \text{for } y=x\,,\\ (1-A_1(x))K_2(y) & \text{for } y>x\,.\end{cases}$$

Theorem 5.6.1 *There is a pure Nash equilibrium in the Sakaguchi silent-noisy duel if and only if*

$$m_1<m_2 \quad \textit{and} \quad x_2^*\geq m_1 \quad \textit{or} \quad m_2<m_1 \quad \textit{and} \quad x_1^*\geq m_2\,. \tag{5.20}$$

(a) If $m_1<m_2$ and $x_2^*\geq m_1$, then $(x_*,\, y_*)$ is a pure Nash equilibrium with the payoff vector $(\eta_1,\,\eta_2)$ if and only if $(x_*,\, y_*)=(m_1,\, m_2)$ and $\eta_1=K_1(m_1), \eta_2=(1-A_1(m_1))K_2(m_2)$.

(b) If $m_2<m_1$ and $m_2\leq x_1^*$, then (x_*,y_*) is a pure Nash equilibrium with the payoff vector (η_1,η_2) if and only if $(x_*,y_*)=(x,m_2)$, where $x\in(m_2,\,1]$ and $\eta_1=(1-A_2(m_2))K_1(m_1), \eta_2=K_2(m_2)$.

Theorem 5.6.2 *Let $x_1^*<m_2\leq m_1$. $(F_1,\,F_2)$ is a Nash equilibrium with the payoff vector $(\eta_1,\,\eta_2)$ in the Sakaguchi silent-noisy duel where*
(a) if $x_1^*\geq a_1(m_2)$ then

$$F_1(x)=\begin{cases}0 & \textit{for } x\in[0,a)\,,\\ \varphi_1(a,x) & \textit{for } x\in[a,m_2)\,,\\ 1 & \textit{for } x\in[m_2,1]\,,\end{cases}$$
$$F_2(x)=\chi_{[a,1]}(x)$$

and

$$\eta_1=(1-A_2(a))K_1(m_1)\,,$$
$$\eta_2=K_2(a)\,,$$

where $a\in[a_1(m_2),\,x_1^*]$,
(b) if $x_1^*<a_1(m_2)$, then let

$$F_1(x)=\begin{cases}0 & \textit{for } x\in[0,a_1(m_2))\,,\\ \varphi_1(a_1(m_2),x) & \textit{for } x\in[a_1(m_2),\,m_2)\,,\\ 1 & \textit{for } x\in[m_2,\,1]\,,\end{cases}$$
$$F_2(x)=\begin{cases}0 & \textit{for } x\in[0,\,a)\,,\\ \psi_2(x,m_2) & \textit{for } x\in[a,\,m_2)\,,\\ 1 & \textit{for } x\in[m_2,\,1]\end{cases}$$

and

$$\eta_1=K_1(a_1(m_2))\,,$$
$$\eta_2=K_1(a_1(m_2))\,.$$

Theorem 5.6.3 *Let $x_2^* < m_1 < m_2$. (F_1, F_2) is a Nash equilibrium with the payoff vector (η_1, η_2) in the Sakaguchi silent-noisy duel where*

$$F_1(x) = \begin{cases} 0 & \text{for } x \in [0, a_*), \\ \varphi_1(a_*, x) & \text{for } x \in [a_*, m_1), \\ \varphi_1(a_*, m_1) & \text{for } x \in [m_1, m_2). \\ 1 & \text{for } x \in [m_2, 1], \end{cases}$$
$$F_2(x) = \begin{cases} 0 & \text{for } x \in [a, m_1), \\ \psi_2(x, a_*) & \text{for } x \in [a, m_1), \\ 1 & \text{for } x \in [m_1, 1] \end{cases}$$

and

$$\eta_1 = K_1(a_*),$$
$$\eta_2 = K_1(a_*),$$

where a_* is the unique root in $(0, m_1)$ of the following equation

$$K_2(m_2)\left(\frac{K_2(a)}{K_2(m_1)} - A_1(m_1)(1 - \varphi_1(a, m_1))\right) = K_2(a).$$

It is clear that Theorems from 5.6.1 to 5.6.3 give Nash equilibria for all possible relations between x_1^*, x_2^*, m_1 and m_2.

5.7 A Duel over a Cake

Consider an example of non-zero-sum duel on an infinitive interval suggested by Hamers [53]. In this game two players divide a cake of size 1. At time $x = 0$ player i has the initial right to receive the positive amount α_i, where $i = 1, 2$ and $\alpha = \alpha_1 + \alpha_2 < 1$. Player 1 and 2 must choose a moment of time $x \in [0, \infty)$ and $y \in [0, \infty)$, respectively. If $x < y$, then player 1 gets the discounted part $\alpha_1 \delta^x$ of the cake, while player 2 receives the discounted remaining $(1 - \alpha_1)\delta^y$ of the cake, where $\delta \in (0, 1)$. If $x > y$, then player 2 gets $\alpha_2 \delta^y$ part of the cake and player 1 receives $(1 - \alpha_2)\delta^x$ part. If $x = y$, then each player receives his discounted initial right and they share the remaining part equally. Then if player 1 and player 2 employ strategies x and y, respectively, their payoffs are given as follows:

$$M_1(x, y) = \begin{cases} \alpha_1 \delta^x & \text{for } x < y, \\ (\alpha_1 + \bar{\alpha}/2)\delta^x & \text{for } x = y, \\ \bar{\alpha}_2 \delta^x & \text{for } x > y, \end{cases}$$
$$M_2(x, y) = \begin{cases} \alpha_2 \delta^y & \text{for } x > y, \\ (\alpha_2 + \bar{\alpha}/2)\delta^x & \text{for } x = y, \\ \bar{\alpha}_1 \delta^y, & \text{for } x < y, \end{cases}$$

where $\bar{\alpha} = 1 - \alpha$. Given a mixed strategy F_2 for player 2 and a pure strategy x for player 1, the payoff to player 1 is given by

$$M_1(x, F_2) = \alpha_1\delta^x F_2(x-0) + \alpha_2\delta^x(1-F_2(x)) + (\alpha_1+\bar{\alpha}/2)\delta^x P_2(x).$$

Thus, by right-continuity of distribution functions F_1 and F_2 for any $x \in [0, \infty)$

$$\begin{aligned} M_2(F_1, x) &= (\alpha_2 + \bar{\alpha}F_1(x) + \bar{\alpha}q_2(x)/2)\delta^x, \\ M_1(x, F_2) &= (\alpha_1 + \bar{\alpha}F_2(x) - \bar{\alpha}q_2(x)/2)\delta^x. \end{aligned} \tag{5.21}$$

5.7.1 Properties of Nash Equilibrium

In this section we obtain properties of the Nash equilibria of the duel over a cake which will enable us to establish our main results in the next section.

Lemma 5.7.1 *Let (F_1, F_2) be a Nash equilibrium with the payoff vector (η_1, η_2) and $\epsilon > 0$ then*

(i) If F_2 has a jump at x then $M_2(F_1, x) = \eta_2$.

(ii) If $M_2(F_1, x) < \eta_2 - \epsilon$ for all $x \in (a, b)$ then $(a, b) \cap \operatorname{supp} F_2 = \emptyset$ and F_2 is constant in $[a, b)$.

(iii) If $x \in \operatorname{supp} F_2$ and $M_2(F_1, x)$ is continuous at x then $M_2(F_1, x) = \eta_1$.

Intuition suggests that, if a player acts with positive probability at time x, then the other player would be better off acting just after x rather than at time x or just before x. As a result, in a Nash equilibrium, we would not expect both players to act with positive probability at the same time. These ideas are formalized in the following result.

Lemma 5.7.2 *(i) For a distribution function F_1 we have*

$$\begin{aligned} M_2(F_1, x+0) &= M_2(F_1, x) + \bar{\alpha}\delta^x q_1(x)/2, \\ M_2(F_1, x-0) &= M_2(F_1, x) - \bar{\alpha}\delta^x q_1(x)/2. \end{aligned}$$

In particular $M_2(F_1, \cdot)$ is continouos at a point if anf only if F_1 is continuous there.

(ii) Let (F_1, F_2) be a Nash equilibrium then $q_1(x)q_2(x) = 0$ for $x \in [0, \infty)$.

If a player intends to act in a time interval in which he knows his opponent is not going to act, common sense tells us that acting early in that interval cannot be worse than acting late in it. This is made more precise in the next lemma.

Lemma 5.7.3 *Let $0 \le a < b$ and the distribution function F_2 is constant in $[a, b)$, then $M_1(\cdot, F_2)$ is a strictly decreasing function in $[a, b)$.*

Proof Let F_2 is constant in for $[a, b)$. Thus, by (5.21), for $x \in (a, b)$ we have, since $q_2(x) = 0$

$$M_1(x, F_2) = (\alpha_1 + \bar{\alpha}F_2(x))\delta^x = (\alpha_1 + \bar{\alpha}F_2(a))\delta^x.$$

Clearly it is strictly decreasing function. The result now follows.

It is now convenient to introduce some extra notation. For a mixed strategy F_i of player i we set (by right continuity)

$$X_i = X_i(F_i) = \inf\{x : F_i(x) = 1\} = \min\{x : F_i(x) = 1\}$$

and call X_i the final action time for player i. Also, let

$$X = \min\{X_1, X_2\}.$$

Our next lemma proves that, in general, there are a time when a player acts with positive probability in a Nash equilibrium. In particular (i) tells us that a player playing against rival will never act with positive probability once the game has started. However (ii) tells us that each player has a positive probability of acting in any interval of time between the start ant the minimum of the players' final action times.

Lemma 5.7.4 *A Nash equilibrium* (F_1, F_2) *with the payoff vector* (η_1, η_2) *has the following properties*

(i) F_i *is a continuous function in* $(0, \infty)$,

(ii) F_i *is strictly increasing in* $[0, X]$ *where* $i = 1, 2$.

Proof Let (F_1, F_2) be a Nash equilibrium with the payoff (η_1, η_2). We will prove all two results by means of a contradiction arguments.

(i) Without loss of generality suppose that F_1 is not continuous at x where $x > 0$. By Lemma 5.7.2(ii) F_2 is continuous at x so $F_2(x-0) = F_2(x)$. By Lemma 5.7.2(i) there is a positive ϵ such that

$$M_2(F_1, x - w) < M_2(F_1, x) - \bar{\alpha}\delta_2^x q_1(x)/4 \quad \text{for} \quad w \in (0, \epsilon).$$

Hence by Lemma 5.7.1(ii), F_2 is constant in $[x-\epsilon, x)$. Thus, by Lemma 5.7.3, $M_1(\cdot, F_2)$ is strictly decreasing in $[x-\epsilon, x)$. By the corresponding result to Lemma 5.7.2(i), $M_1(\cdot, F_2)$ is continuous at x because F_2 is. So,

$$M_1(x, F_2) < M_1(x - \epsilon, F_2) \le \eta_1.$$

Since F_1 has a discontinuity at x we have a contradiction to Lemma 5.7.1(i). So, (i) is established.

(ii) Without loss of generality suppose that F_2 is not strictly increasing in $[0, X]$ then there are non-negative x_1 and x_2 satisfying $x_1 < x_2 \le X$ such that $F_2(x_1) = F_2(x_2)$. Since $x_1 < X$, $F_2(x_1) < 1$. Put

$$a = \inf\{x : F_2(x) = F_2(x_1)\} \quad \text{and} \quad b = \sup\{x : F_2(x) = F_2(x_1)\},$$

then $a < b$ and $b \in \operatorname{supp} F_2$. By Lemma 5.7.3 $M_1(\cdot, F_2)$ is streactly decreasing in $[a, b)$. On taking $\delta = (b-a)/2 > 0$, it follows from corresponding result to Lemma 5.7.1(ii) that F_1 is constant in $[a+\delta, b)$. Since $a+\delta < b \leq X$, $F_1(x) < 1$ on $[a+\delta, b)$ so, by the corresponding result to Lemma 5.7.3, $M_2(F_1, \cdot)$ is strictly decreasing in $[a+\delta, b)$.

By (i) F_1 and F_2 are continuous at b. Since

$$M_1(a+\delta, F_2) > M_1(b-\delta, F_2) \quad \text{and} \quad M_2(F_1, a+\delta) > M_2(F_1, b-\delta),$$

it follows from Lemma 5.7.2(i) that there is an $\epsilon > 0$ such that, for $x \in (b-\epsilon, b+\epsilon)$, $M_1(x, F_2)$ is bounded away from η_1 or $M_2(F_1, x)$ is bounded away from η_2. However, by Lemma 5.7.1(ii), the latter would contradict that $b \in \operatorname{supp} F_2$ so the former holds and F_1 is constant in $[b-\epsilon, b+\epsilon)$ by the corresponding result to Lemma 5.7.1(ii). Thus, by Lemma 5.7.3, $M_2(F_1, x)$ is bounded away from η_2 in $[b-\epsilon/2, b+\epsilon/2)$ and, by Lemma 5.7.1(ii), $b \notin \operatorname{supp} F_2$ which is a contradiction. This completes the proof of Lemma 5.7.4.

The next result deals with the action times in a Nash equilibrium. It will be wshowen that the final action times of the players are equil.

Lemma 5.7.5 *Let (F_1, F_2) be a Nash equilibrium, then $X_1 = X_2$.*

Proof Let (F_1, F_2) be a Nash equilibrium. Suppose that the assertion is false. Say, $X_2 > X_1$, then X_1 is finite. Let $\eta > 0$, then, for $x \geq X_1 + \eta$, we have from (5.21)

$$\begin{aligned} M_2(F_1, x) - M_2(F_1, X_1 + \eta/2) &= \bar{\alpha}_1(\delta^x - \delta^{X_1+\eta/2}) \\ &\leq \bar{\alpha}_1(\delta^{X_1+\eta} - \delta^{X_1+\eta/2}) < 0. \end{aligned}$$

It follows from Lemma 5.7.1(ii) that $F_2(x)$ is constant for $x \geq X_1 + \eta$ so $F_2(x) = 1$ for $x \geq X_2 + \eta$ since $\lim_{x\to\infty} F_2(x) = 1$. Since $\eta > 0$ is arbitrary, $X_2 \leq X_1$ and we have a contradiction.

We now show that only strategies of a very particular form can appear in a Nash equilibrium.

Lemma 5.7.6 *If (F_1, F_2) is a Nash equilibrium then, for $i = 1, 2$*

$$F_i(x) = \begin{cases} \dfrac{\eta_j - \alpha_j\delta^x}{\bar{\alpha}\delta^x} & \text{for } x \in [0, X), \\ 1 & \text{for } x \in [X, \infty). \end{cases} \tag{5.22}$$

Also,

$$X = \ln_\delta(\eta_1/\bar{\alpha}_2) = \ln_\delta(\eta_2/\bar{\alpha}_1) \tag{5.23}$$

and

$$\eta_i \geq \alpha_i \quad \text{for} \quad i = 1, 2. \tag{5.24}$$

Proof Let (F_1, F_2) be a Nash equilibrium with the payoff vector (η_1, η_2) and suppose $X = X_1$. If $X = 0$ there is nothing to prove so assume $X > 0$. Now F_1 and F_2 are continuous in $(0, X)$ by Lemma 5.7.4(i) and (ii). Further they are strictly increasing in $[0, X]$ by Lemma 5.7.4(iii) so in $[0, X]$ is in the support of both F_1 and F_2. By Lemma 5.7.2(i) $M_1(\cdot, F_2)$ and $M_2(F_1, \cdot)$ are continuous in $(0, X)$ so, by Lemma 5.7.1(iii), they are constant on $(0, X)$ with values η_1 and η_2 respectively. Now, for $x \in (0, X)$ we have

$$\eta_1 = M_1(x, F_2) = (\alpha_1 + \bar{\alpha} F_2(x))\delta^x$$

and

$$F_2(X) = 1, \quad F_2(0) \geq 0\,.$$

Now the result follows.

Theorem 5.7.1 *In the game over a cake there is unique Nash equilibrium* (F_1, F_2) *with the payoff vector* (η_1, η_2). *Also,* F_i *for* $i = 1, 2$ *are given by (5.22) and*

$$\eta_i = \begin{cases} \dfrac{\alpha_j \bar{\alpha}_j}{\bar{\alpha}_i} & \text{if } \alpha_i < \alpha_j\,, \\ \alpha_i & \text{if } \alpha_i \geq \alpha_j\,, \end{cases} \tag{5.25}$$

where $i = 3 - j$, $j = 1, 2$.

Proof Let (F_1, F_2) be a Nash equilibrium with the payoff vector (η_1, η_2). Then, by Lemma 5.7.6, (F_1, F_2) is given (5.22). By Lemma 5.7.2(ii), $q_1(0)q_2(0) = 0$. Thus, by (5.22) and (5.23) either

$$\eta_1 = \alpha_1, \quad \eta_2 = \alpha_1\bar{\alpha}_1/\bar{\alpha}_2 \tag{5.26}$$

or

$$\eta_1 = \alpha_2\bar{\alpha}_2/\bar{\alpha}_1, \quad \eta_2 = \alpha_2\,. \tag{5.27}$$

Note that

$$\alpha_1\bar{\alpha}_1 - \alpha_2\bar{\alpha}_2 = (\alpha_1 - \alpha_2)\bar{\alpha}\,.$$

Thus, if $\alpha_1 > \alpha_2$ (5.24) holds for (5.26) and does not hold for (5.27). Similarly, if $\alpha_1 < \alpha_2$ (5.24) holds for (5.27) and does not hold for (5.26). So, if a Nash equilibrium exists it is unique and given by (5.22) with the payoff vector given by (5.25). It is clear that it is Nash equilibrium.

Corollary 5.7.1 *Let* $\alpha_1 = \alpha_2 = \sigma$ *then in the game over a cake there is unique Nash equilibrium* (F, F) *with the payoff vector* (σ, σ). *Also,*

$$F(x) = \begin{cases} \sigma(\delta^{-x} - 1)/\bar{\sigma} & \text{for } x \in [0, \ln_\delta(\sigma/\bar{\sigma})\,, \\ 1 & \text{for } x \in [\ln_\delta(\sigma/\bar{\sigma}), \infty)\,. \end{cases}$$

5.7.2 A Generalization of the Duel over a Cake

Consider a generalization of the game over a cake (Baston and Garnaev [18]), where player i has discount factor δ_i and becomes informed of his opponent's choice with probability $\bar{\beta}_j$, where $\beta_i \in [0,1]$ and $j = 3-i$, $i = 1, 2$. Then if player 1 and player 2 employ strategies x and y, respectively, their payoffs are given as follows

$$M_1(x,y) = \begin{cases} \alpha_1\delta_1^x & \text{for } x < y\,, \\ (\alpha_1 + \bar{\alpha}/2)\delta_1^x & \text{for } x = y\,, \\ \bar{\alpha}_2(\bar{\beta}_2\delta_1^y + \beta_2\delta_1^x) & \text{for } x > y\,, \end{cases}$$

$$M_2(x,y) = \begin{cases} \alpha_2\delta_2^y & \text{for } x > y\,, \\ (\alpha_2 + \bar{\alpha}/2)\delta_2^y & \text{for } x = y\,, \\ \bar{\alpha}_1(\bar{\beta}_1\delta_2^x + \beta_1\delta_2^y) & \text{for } x < y\,. \end{cases}$$

We now use the properties to characterize the Nash equilibria when both players are non-noisy and the pure Nash equilibria when at least one of the players is noisy. To do this we use the following notation.

Let $\beta_1\beta_2 > 0$ and

$$T_i = \begin{cases} -\dfrac{\bar{\alpha}\ln(\bar{\alpha}_i\beta_j/\alpha_j)}{\ln(\delta_j)\gamma_i} & \text{for } \gamma_i \neq 0\,, \\ -\dfrac{\bar{\beta}_j}{\ln(\delta_j)\beta_j} & \text{for } \gamma_i = 0\,, \end{cases}$$

and

$$F_{*i}(x) = \begin{cases} \alpha_j\{\delta_j^{-\gamma_i(x+T_i-T)/\bar{\alpha}} - 1\}/\gamma_i & \text{for } \gamma_i \neq 0\,, \\ (x + T_i - T)/T_i, & \text{for } \gamma_i = 0\,, \end{cases}$$

where $x \in [0, T]$ and

$$F_{*i}(x) = 1 \text{ for } x \in (T, \infty)\,,$$

where $T = \min\{T_1, T_2\}$ and $i = 3-j$, $j = 1, 2$.

Note that $\bar{\alpha}_i\beta_i/\alpha_j$ is greater or less than 1 according as γ_i greater than or less than 0 so $T_i > 0 (i = 1, 2)$. Further F_{*i} is a probability distribution function with support $[0, T]$.

Theorem 5.7.2 *If both players are not noisy (so, $\beta_1\beta_2 > 0$) then (F_{*1}, F_{*2}) is unique Nash equilibrium with the payoff vector (η_1, η_2), where $\eta_i = \alpha_i + \bar{\alpha}F_{*j}(0), i = 3-j, j = 1, 2$.*

If at least one of the players is noisy the uniqueness of Nash equilibrium does not hold. The next theorem deals with the case when one of the player is noisy and the other non-noisy. In this case there are both pure equilibrium and Nash equilibria in which the non-noisy player's strategy is a continuous distribution function. Also, the noisy player in this plot always acts immediately.

Theorem 5.7.3 *Let one player (say, player 2) is noisy and the other one is not noisy. So, $\beta_1 = 0$ and $\beta_2 > 0$. Then*
(a) the set of pure Nash equilibria is given by

$$\left\{(0,\tau) : \tau \geq \frac{\ln(\alpha_1/\bar{\alpha}_2)}{\ln(\delta_1)}\right\},$$

with the payoff vector $(\alpha_1, \bar{\alpha}_1)$,
(b) if

$$F(x) = \begin{cases} f(x) & \text{for } x \in [0,T], \\ 1 & \text{for } x \in (T,\infty), \end{cases}$$

where

$$f(x) = \begin{cases} \alpha_1 \dfrac{\delta_1^{-\gamma_2 x/\bar{\alpha}-1}}{\gamma_2} & \text{for } \gamma_2 \neq 0, \\ \dfrac{-\ln(\delta_1)\beta_2 x}{\beta_2} & \text{for } \gamma_2 = 0 \end{cases}$$

and T is the unique root of the equation $f(x) = 1$, then $(0, F)$ is a Nash equilibrium with the payoff vector $(\alpha_1, \bar{\alpha}_1)$.

Finally we consider the case when both players are noisy. In this case there are many Nash equilibria including one in which neither player acts with probability 1 at the start.

Theorem 5.7.4 *Let both players are noisy. So, $\beta_1 = \beta_2 = 0$. Then*
(a) the set of pure Nash equilibria is given by

$$\left\{(0,\tau) : \tau \geq \frac{\ln(\alpha_1/\bar{\alpha}_2)}{\ln(\delta_1)}\right\} \cup \left\{(\tau,0) : \tau \geq \frac{\ln(\alpha_2/\bar{\alpha}_1)}{\ln(\delta_2)}\right\},$$

with payoff vectors $(\alpha_1, \bar{\alpha}_1)$ and $(\bar{\alpha}_2, \alpha_2)$, respectively.
(b) $(0, F_2)$ and $(F_1, 0)$ are Nash equilibria with payoff vectors $(\alpha_1, \bar{\alpha}_1)$ and $(\bar{\alpha}_2, \alpha_2)$, respectively, where $F_i(x) = 1 - \delta_j^{\alpha_j x/\bar{\alpha}}$ and $i = 3 - j$, $j = 1, 2$.

5.8 R & D Game

In a paper presented at a UK-Japanese workshop on Stochastic Modelling in Innovative Manufacturing held at Cambridge in July 1995, Teraoka and Yamada [106] modelled the strategic aspects of production development in manufacturing by means of a game. In this game two players compete to dominate the market of a new product (for example, a new mainboard for computers) by putting their product on the market after undertaking research and development (R & D) on it; the player who spends more time on R & D comes to dominate the market and is the only one to make a profit.

For their model Teraoka and Yamada make the following assumptions concerning the two firms.

(i) The firms are equal in every respect so that, in particular, the costs for R & D are the same for both.

(ii) There is no initial fixed charge for setting up R & D.

(iii) Only the firm which spends the longest time on R & D earns any revenue.

(iv) Firms can spend an indefinitely large amount of time on R & D.

With these assumptions their model is a very particular game of timing with complete information called a war of attrition.

Baston and Garnaev [19] investigated a variation of the Teraoka-Yamada model which does not require the above assumptions. In this section we consider a particular case of their model. Like Teraoka and Yamada, assume the firms can start marketing the product at the same time, namely $t = 0$. Clearly, if a product can be put on the market at time $t = 0$, some R & D must have been undertaken already. However extra costs such as the buying of more specialized equipment or the funding of expanded facilities may be required if further development and testing of the product is to be undertaken after time 0. We will therefore introduce a fixed non-negative setting-up cost for a firm that does not market its product at time $t = 0$. This cost is in addition to a running cost per unit time for each firm, where different firms can have different running costs. It is also reasonable to expect a product to have a limited life-span before it becomes obsolete or subject to intense competition; by suitable choice of units we can assume this life-span is 1.

As a firm learns when its competitor puts its product on the market, it might appear natural to treat the situation as a noisy game. However a traditional noisy duel setting would lead to a firm marketing its product immediately it learns that its rival has. If a firm has planned its R & D, it might not be possible for it to react rapidly to a move by its competitor so a silent duel scenario may be more appropriate and this is the stance we shall adopt (Teraoka and Yamada consider both noisy and silent settings). Thus a (pure) strategy for a firm will be a time t representing the time it markets its product; we will suppose that the longer the time, the more R & D there has been done and so the better the product.

As the firm putting its product on the market first will initially have a monopoly of the market, we will assume that it makes a profit of A per unit time until its rival enters the market. At that point the entering firm has the better product because it has done more R & D and so we will assume that it takes over the market earning a profit B per unit time until time 1 while the first player in the market gets no revenue during that period. The latter is merely a simplifying assumption as the analysis can easily be amended to cope with the firm earning a profit of A' per unit time.

A natural solution concept for these games is that of Nash equilibria (provided there are comparatively few of them). Our results show that there is always a unique Nash equilibrium when the running costs of the firm are the same. However when they differ, a Nash equilibrium (again unique) only exists if the setting up cost is either zero or high compared to the profit rate B.

5.8.1 Formulation of the Game

We now provide a mathematical formulation of our model. It is a non-zero sum two-person game in which each player chooses a number from the interval $[0, 1]$. Since the decision to be taken is when to market the product in $[0, 1]$, it is natural to consider only the costs that are incurred after time 0. Thus, for each player, the costs are 0 if the product is put on the market immediately and $C_i\tau + S$ if the product is marketed at time τ, where S is the setting-up cost and C_i is the running cost per unit time for player i. If one player markets his product at time t and the other at time $\tau > t$, then the first player gets a profit of $A(\tau - t)$ while the other gets a profit of $B(1 - \tau)$. Of course there is the possibility that the firms will decide to put their products on the market at the same time. Under our assumptions, the products will be of the same quality so we can expect them to yield the same profit D per unit time. Suppose the two players act at the same time, then the market yields a profit of $2D$ per unit time whereas, if one acts and then the other acts almost immediately afterwards, the market effectively yields a profit of B per unit time. It therefore appears reasonable to expect that $2D$ and B should be approximately equal, perhaps even $d = B/2$. However a firm may charge one price if it thinks it is going to be first in the market but another if it knows that it is second. Hence a weaker assumption than $D = B/2$ seems preferable and we will assume only that $D < B$. Taking a player's payoff to be the profit minus the R & D expenditure, the payoff M_i to player i when players 1 and 2 choose pure strategies x_1 and x_2 respectively is given by

$$M_i(x_1, x_2) = \begin{cases} A(x_j - x_i) - C_i x_i - S & \text{for } x_i < x_j \text{ and } x_i > 0\,, \\ D(1 - x_i) - C_i x_i - S & \text{for } x_i = x_j \text{ and } x_i > 0\,, \\ B(1 - x_i) - C_i x_i - S & \text{for } x_i > x_j \text{ and } x_i > 0\,, \\ A x_j & \text{for } x_j > 0 \text{ and } x_i = 0\,, \\ D & \text{for } x_j = 0 \text{ and } x_i = 0\,, \end{cases}$$

where $j = 3 - i$ and $i = 1, 2$.

As we shall see, Nash equilibria do exist under some circumstances but, in general, mixed strategies are required. In this context a mixed strategy is a cumulative probability distribution function F in $[0, 1]$.

It is clear from the form of the payoffs that, to every result for player 1, there is a corresponding result for player 2 and vice-versa. To avoid repetition we will state a result for one player and take the other for granted.

5.8.2 Auxilary Results

In this section we obtain properties of the Nash equilibria (F_1, F_2) which will enable us to establish our main results in Section 18.4. In particular we show that the F_i are continuous in (0, 1) and that, when $S > 0$, they have the same probability of putting the product on the market immediately.

For a given mixed strategy F_2 of player 2 and a pure strategy x of player 1, the payoff to player 1 is given by

$$\begin{aligned} M_1(x, F_2) &= \int_{[0,x)} (B - S - (B + C_1)x)\, dF_2(y) \\ &\quad + \int_{(x,1]} (Ay - (A + C_1)x - S)\, dF_2(y) \\ &\quad + (D - S - (D + C_1)x)q_2(x) \quad \text{for} \quad x > 0\,, \\ M_1(0, F_2) &= Dq_2(0) + \int_{(0,1]} Ay\, dF_2(y)\,, \end{aligned} \tag{5.28}$$

where

$$q_i(x) = F_i(x) - F_i(x - 0) \quad \text{for} \quad i = 1, 2\,.$$

Since F_2 is a right-continuous function, from (5.28) we have

$$\begin{aligned} M_1(0 + 0, F_2) &= M_1(0, F_2) + (B - D)q_2(0) - S\,, \\ M_1(x + 0, F_2) &= M_1(x, F_2) \\ &\quad + (B - D)(1 - x)q_2(x) \quad \text{for} \quad x \in (0, 1)\,, \\ M_1(x - 0, F_2) &= M_1(x, F_2) - D(1 - x)q_2(x) \quad \text{for} \quad x \in (0, 1]\,. \end{aligned} \tag{5.29}$$

The corresponding relations to (5.28) and (5.29) hold for $M_2(F_1, y)$.

We require the following property which follows from the definitions of support and Nash equilibrium.

Remark 5.8.1 *If* (F_1, F_2) *is a Nash equilibrium with payoff vector* (η_1, η_2), *then* $M_1(x, F_2) = \eta_1$ *if one of the following conditions hold:*
(i) $x \in \operatorname{supp} F_1$ *and* $M_1(t, F_2)$ *is continuous at* $t = x$,
(ii) $q_1(x) > 0$.

If it is known that a firm will not put its product on the market during the time interval I, then, it is intuitively obvious that, if the other firm puts its product on the market during the time interval I, it should do so as early as possible in I. This statement is made precise in the next lemma.

Lemma 5.8.1 *Let $(a, b] \subset (0, 1]$ and F_2 be constant in $(a, b]$, then $M_1(\cdot, F_2)$ is strictly decreasing in $(a, b]$.*

Proof Let $x_1, x_2 \in (a, b]$ with $x_1 < x_2$, then, since F_2 is constant in $(a, b]$ we obtain

$$M_1(x_2, F_2) - M_1(x_1, F_2) = -(x_2 - x_1) \times \left((A + C_1) \int_{(x_2,1]} dF_2(y) + (B + C_1) \int_{[0,x_1)} dF_2(y) \right) < 0 .$$

This completes the proof of Lemma 5.8.1.

Our next result shows that, in a Nash equilibrium (F_1, F_2), the F_i $(i = 1, 2)$ have a very simple structure. They are continuous except possibly at 0; furthermore the period over which a firm may put its product on the market is the same for each firm. It should be noted that the proofs of the next two lemmas depend only on (5.29) and Lemma 5.8.1, not on the particular form of M_1 given in (5.28).

Lemma 5.8.2 *Let (F_1, F_2) be a Nash equilibrium with the payoff vector (η_1, η_2), then*
(a) if $(a, b] \subset (0, 1)$ and F_j is constant in $(a, b]$ then F_i is constant in $(a, b]$,
(b) F_i $(i = 1, 2)$, $M_1(\cdot, F_2)$ and $M_2(F_1, \cdot)$ are continuous in (0, 1],
(c) $\operatorname{supp} F_i = [0, X]$ $(i = 1, 2)$, where $X = \min\{X_1, X_2\}$ and $X_i = \min\{x : F_i(x) = 1\}$.

Proof (a) Let $i = 1$ then, by Lemma 5.8.1, $M_1(x, F_2)$ is strictly decreasing in $(a, b]$. Hence $(a, b] \cap \operatorname{supp} F_1 = \emptyset$. This completes the proof of (a).

(b) First note that, for any F, $M_2(F, 1) = -C_2 - S < M_2(F, 0) \le \eta_2$ so $q_2(1) = 0$ and F_2 is continuous at 1. Now suppose that F_2 is not continuous at $x \in (0, 1)$, then $q_2(x) > 0$. By (5.29), we have that

$$M_1(x + 0, F_2) - M_1(x - 0, F_2) = B(1 - x)q_2(x) > 0 ,$$

$$M_1(x + 0, F_2) - M_1(x, F_2) = (B - D)(1 - x)q_2(x) > 0$$

so there is a sufficiently small positive ω such that

$$M_1(z_2, F_2) > M_1(z_1, F_2) \quad \text{for any} \quad z_2 \in (x, x + \omega], \ z_1 \in [x - \omega, x] .$$

Thus $(x - \omega, x) \cap \operatorname{supp} F_1 = \emptyset$ and $q_1(x) = 0$. Therefore F_1 is constant in $[x - \omega, x]$. Then, by (a), F_2 is constant in $(x - \omega, x]$ and, since it is right-continuous, F_2 is also continuous at x and we have a contradiction. Hence F_1 and F_2 are continuous in $(0, 1]$. By (5.29), $M_1(\cdot, F_2)$ and $M_2(F_1, \cdot)$ are continuous in $(0, 1]$ and the proof of (b) follows.

(c) It is immediate from (a) that $F_i(x) = 1$ for $x \geq X$. Suppose that F_2 is not strictly increasing in $(0, X)$, then there are $y_1 < y_2$ such that $F_2(y_1) = F_2(y_2) < 1$. Let $b = \sup\{y : F_2(y) = F_2(y_1)\}$ then $b \in \operatorname{supp} F_2$ and, by (a), F_1 is constant in (y_1, b) and $b \in \operatorname{supp} F_1$. Further, by (b), F_2 is continuous at $y = b$. Thus, by Remark 5.8.1, $M_1(b, F_2) = \eta_1$. But, by Lemma 5.8.1, $M_1(x, F_2)$ is strictly decreasing in $(y_1, b]$ so $M_1(b, F_2) < \eta_1$ and (c) follows.

We will now show that, if there is a setting-up cost for R & D, then, in any Nash equilibrium, each firm has a positive probability of putting its product on the market at time $t = 0$. This probability can take at most two values and is the same for each firm; if the cost is large compared with the prospective profit B, both firms will market their product immediately.

Lemma 5.8.3 *Let $S > 0$ and $(F_1,\ F_2)$ be a Nash equilibrium with the payoff vector $(\eta_1,\ \eta_2)$ then $q_1(0) = q_2(0) = \alpha$ where $\alpha \in \{1,\ S/(B - D)\}$.*

Proof If $q_1(0) = 1$, then $q_2(0) = 1$ by Lemma 5.8.2(c). If $0 \leq q_1(0) < 1$, then $\operatorname{supp} F_1 = [0,\ X]$ for some $X > 0$ by Lemma 5.8.2(c) so, by Remark 5.8.1, $M_1(x, F_2) = \eta_1$ for $x \in (0, X)$. Now $q_1(0) \neq 0$, for, if $q_1(0) = 0$, (5.29) gives $M_1(0, F_2) > M_1(0 + 0, F_2) = \eta_1$ which is impossible. Thus $M_1(0, F_2) = \eta_1$ and (5.29) now requires $q_2(0) = S/(B - D) > 0$ because $S > 0$ and $B > D$. The lemma now follows for, if $S/(B - D) \leq 1$, the above argument on q_2 gives $q_1(0) = S/(B - D) = q_2(0)$.

If $(F_1,\ F_2)$ is to be a Nash equilibrium, we can determine the form the F_i $(i = 1,\ 2)$ must take in $[0,\ X)$ where $[0,\ X]$ is the period (see Lemma 5.8.2(c)) over which the firms may act. The function H defined below plays a major role.

$$H_i(x, a) = \begin{cases} \dfrac{A + C_j}{B - A}\left((1 - x)^{\frac{A-B}{B}} - 1\right) + a(1 - x)^{\frac{A-B}{B}} & \text{for } A \neq B\,, \\ -(A + C_j)\ln(1 - x)/A + a & \text{for } A = B\,, \end{cases}$$

where $j = 3 - i$ and $i = 1,\ 2$.

Remark 5.8.2 *Note that $H_i(0, a) = a$, H_i is strictly increasing in $[0,\ 1)$ and that, as $x \to 1 - 0$, $H_i \to \infty$ if $B > A$ and $H_i \to (A + C_i)/(A - B) > 1$ if $A > B$. Thus, for $a \leq 1$, there is a unique $x \in (0,\ 1)$ such that $H_i(x, a) = 1$.*

Lemma 5.8.4 *Let $(F_1,\ F_2)$ be a Nash equilibrium then F_i is of the form*

$$F_i(x) = H_i(x, q_i(0)) \quad \textit{for} \quad x \in [0,\ X),$$

for some $q_i(0) \in [0,\ 1]$ and some $X \in [0,\ 1)$.

Proof If (F_1, F_2) is a Nash equilibrium with the payoff vector (η_1, η_2), then by Lemma 5.8.2 F_2 consists of a mass part $q_2(0)$ at $x = 0$, a density part over the interval $(0, X)$ and equals 1 for $x \geq X$. Also F_2 is continuous at $x = X$. Thus, we have that

$$M_1(x, F_2) = \eta_1 \quad \text{for} \quad x \in (0, X).$$

Using (5.28) and differentiating yield

$$F_2' = \frac{A + C_1}{B(1-x)} + \frac{B - A}{B(1-x)} F_2 \tag{5.30}$$

with the initial condition

$$F_2(0) = q_2(0).$$

Hence,

$$F_2(x) = H_2(x, q_2(0)) \quad \text{for} \quad x \in [0, X).$$

This completes the proof of Lemma 5.8.4.

The next result provides us with an easy to apply test for checking whether (F_1, F_2) is a Nash equilibrium.

Lemma 5.8.5 *Suppose there are $a_1, a_2 \in [0, 1)$ and $x_* \in (0, 1)$ such that $H_1(x_*, a_1) = 1 = H_2(x_*, a_2)$, then (F_1, F_2) given by*

$$F_i(x) = \begin{cases} H_i(x, a_i) & \text{if } x \leq x_*, \\ 1 & \text{if } x > x_* \end{cases}$$

is a Nash equilibrium if and only if

$$M_1(0, F_2) \leq M_1(0 + 0, F_2) \quad \text{with equality when } a_1 > 0,$$

$$M_2(F_1, 0) \leq M_2(F_1, 0 + 0) \quad \text{with equality when } a_2 > 0.$$

Proof Clearly F_1 and F_2 are cumulative distribution functions on $[0, 1]$. By the way H is defined, $M_1(\cdot, F_2)$ and $M_2(F_1, \cdot)$ are constant on $(0, x_*]$. Hence for a Nash equilibrium, we need

$$M_1(0, F_2) \leq M_1(0 + 0, F_2) \quad \text{and} \quad M_1(x, F_2) \leq M_1(0 + 0, F_2) \quad \text{for } x > x_*$$

and similar inequalities for M_2. The second inequality always holds by Lemmas 5.8.1 and 5.8.2(b) and the result follows.

5.8.3 The Main Results

In this section we show that, when there is no setting-up cost or the cost is high compared with the profit B, there is always a unique Nash equilibrium. For other values of the setting-up cost there is only a Nash equilibrium when the running costs of the firms are the same.

Our first result makes precise the intuitively obvious statement that the firms will put their product on the market immediately if the setting-up cost is too high.

Theorem 5.8.1 *Let $B - D \le S$ then there is a unique Nash equilibrium (I_0, I_0) where*

$$I_0(x) = \begin{cases} 1 & \text{if } x \ge 0\,, \\ 0 & \text{if } x < 0\,. \end{cases}$$

Proof It is clear that (I_0, I_0) is a Nash equilibrium. Since $S/(B-D) \ge 1$, the uniqueness follows from Lemma 5.8.3.

Theorem 5.8.2 *Let $S = 0$. Then there is the unique Nash equilibrium (F_1, F_2) with the payoff vector (η_1, η_2), and*

$$\begin{aligned} F_i(x) &= \begin{cases} H_i(x, q_i(0)) & \text{for } x \le x_*\,, \\ 1 & \text{for } x > x_*\,, \end{cases} \\ \eta_i &= Bq_j(0) + \int_{(0,1]} Ay\, dF_j(y), \end{aligned} \tag{5.31}$$

where $x_* = \min\{x_{*1}, x_{*2}\}$,

$$x_{*i} = \begin{cases} 1 - \left(\dfrac{A + C_j}{B + C_j}\right)^{\frac{B}{B-A}} & \text{for } A \ne B\,, \\ 1 - \exp(-A/(A + C_j)) & \text{for } A = B\,, \end{cases}$$

for $j = 3 - i$, $i = 1, 2$, and if $x_* = x_{*i}$ then

$$q_{i0} = 0\,,$$

$$q_{j0} = \begin{cases} \dfrac{B + C_i}{B - A}(1 - x_{i*})^{\frac{B-A}{B}} - \dfrac{A + C_i}{B - A} & \text{for } A \ne B\,, \\ 1 + \dfrac{A + C_i}{A} \ln(1 - x_{*i}) & \text{for } A = B\,, \end{cases}$$

where $\{i, j\} = \{1, 2\}$.

Proof Suppose (F_1, F_2) is a Nash equilibrium, then, by (5.29) and Remark 2.1, $q_i(0) > 0$ implies $q_{3-i}(0) = 0$ so $q_1(0)q_2(0) = 0$. Thus keeping in mind that x_{*i} is the unique root in $(0, 1)$ of the equation $H_i(x, 0) = 1$ where $i = 1, 2$ which exists by Remark 5.8.1, the theorem follows from (5.29) and Lemmas 5.8.4 and 5.8.5.

Corollary 5.8.1 *If $S = 0$ then $x_{*1} \le x_{*2}$ if and only if $C_1 \le C_2$. So, increasing price of research makes player produce less high-quility product and put it on the market faster.*

Theorem 5.8.3 *Let $B - D > S > 0$ and $C_1 \ne C_2$ then there is no Nash equilibrium.*

Proof Let a Nash equilibrium (F_1, F_2) exist, then, by Lemma 5.8.4, it is given by (5.31). Since $B - D > S$, then, by (5.29), $q_i(0) < 1$. But, for $C_1 \ne C_2$, the equations $H_i(x, 0) = 1$ for $i = 1, 2$ have different roots so $q_1(0) \ne q_2(0)$ which contradicts Lemma 5.8.3 and the result follows.

Theorem 5.8.4 *Let $B - D > S > 0$ and $C_1 = C_2$ then there is a unique Nash equilibrium (F, F) with payoff vector (η, η) where*

$$F(x) = \begin{cases} H_1(x, S/(B-D)) & \text{for } x \le x^*, \\ 1 & \text{for } x > x^*, \end{cases}$$

$$\eta = A \int_{(0,1]} y\,dF(y) + SD/(B-D),$$

where x^* is the root of the equation $H_1(x, S/(B-D)) = 1$.

Proof The theorem follows from Lemmas 5.8.3-5.8.5 together with (5.29).

5.8.4 Disscussion of the Results

Notice that, if the firm in the market first obtains a profit of A' per unit time where $A' < B$ after its rival enters the market, the expression for $M_1(x, F_2)$ in (5.28) has an additional term $\int_{(x,1]} A'(1-y)dF_2(y)$. However, with this additional term, Lemma 5.8.1 remain unaltered. It is easy to see that so do Lemmas 5.8.2 and 5.8.3. In Lemma 5.8.4 the only change in the differential equation (5.30) is that $B - A'$ replaces B in the denominator. Hence the analysis copes with this situation.

It is interesting to note that assumption that a player can be noisy (so, his opponent can get information of his research, for example, by means of industrial espionage, and apply them at once to put on the market a more high-quility product) leads to pure Nash equilibria. For example, consider the game where both players are noisy. Then their payoffs are given as follows

$$M_i(x_1, x_2) = \begin{cases} -C_i x_i - S & \text{for } x_i < x_j \text{ and } x_i > 0, \\ D(1-x_i) - C_i x_i - S & \text{for } x_i = x_j \text{ and } x_i > 0, \\ B(1-x_j) - C_i x_j - S & \text{for } x_i > x_j \text{ and } x_i > 0, \\ 0 & \text{for } x_i = 0, \end{cases}$$

where $j = 3 - i$, $i = 1, 2$.
Then

(a) if $B \geq S$ then (0, 1) and (1, 0) are Nash equilibria,
(b) if $B \leq S$ then (0, 0) is Nash equilibrium.

Thus, threat to lose the result of research can make a player do not invest money in research. The same occuries in the Teraoka and Yamada game [106]. In the noisy Teraoka and Yamada game the payoffs of the players are given as follows

$$M_i(x_1, x_2) = \begin{cases} -x_i & \text{for } x_i < x_j\,, \\ V(x_i)/2 - x_i & \text{for } x_i = x_j\,, \\ V(x_j) - x_j & \text{for } x_i > x_j\,, \end{cases}$$

where $j = 3 - i$, $i = 1, 2$, $x_1, x_2 \in [0, \infty)$ and $V(x)$ is non-increasing, continuous and non-negative function in $[0, \infty)$ such that $V(0) > 0$. Let $r = \inf\{x : V(x) = 0\}$ then $(0, r)$ and $(r, 0)$ are Nash equilibria with the payoff vector $(-r, -r)$.

In the silent Teraoka and Yamada game the payoffs of the players are given as follows

$$M_i(x_1, x_2) = \begin{cases} -x_i & \text{for } x_i < x_j\,, \\ V(x_i)/2 - x_i & \text{for } x_i = x_j\,, \\ V(x_i) - x_i & \text{for } x_i > x_j\,, \end{cases}$$

where $j = 3 - i$, $i = 1, ts2$. This game has a unique Nash equilibrium (F, F) with the payoff vector (0, 0) where

$$F(x) = \begin{cases} x/V(x) & \text{for } x \in [0, x_*]\,, \\ 1 & \text{for } x \geq x_*\,, \end{cases}$$

where x_* is the unique root of the equation $V(x) = x$.

This game are close to a game of war of attrition by Maynard Smith [72] describing competition berween animals. It consists of displays by the animals, in which victory goes to thc rival which displays longest. The prize consists of control of a particular territory or female, and the disadvantage in such display is the time wasted by both rivals, which could be used more profitable. Say, that the winner receives a payoff of value V and both receive a penalty equil to the lenght of the contest. Assume that there are two contestants: player 1 and 2. The set of pure strategies of a player for this game is $[0, \infty)$. Then, the payoffs of the players are given as follows

$$M_i(x_1, x_2) = \begin{cases} V - x_j & \text{for } x_i > x_j\,, \\ V/2 - x_i & \text{for } x_i = x_j\,, \\ -x_i & \text{for } x_i < x_j\,, \end{cases}$$

where $j = 3 - i$, $i = 1, 2$. This game has a unique Nash equilibrium (F, F) with the payoff vector $(0, 0)$ where

$$F(x) = 1 - \exp(-x/V)\,.$$

Another interesting non-zero-sum game is the one where two players (say, 1 and 2) compete to predict the realized value of a random variable t with uniform distribution in [0, 1]. The player, who has predicted the value not greater than t and nearest to t gets 1 and loser gets 0. If they both predict the same value which is not greater that t each of them gets 1/2. The set of pure strategies of a player for this game is [0, 1]. Then, the payoffs of the players are given by

$$M_i(x_1, x_2) = \begin{cases} x_j - x_i & \text{for } x_j > x_i\,, \\ (1 - x_i)/2 & \text{for } x_i = x_j\,, \\ 1 - x_i & \text{for } x_i > x_j\,, \end{cases}$$

where $j = 3 - i$, $i = 1, 2$. This game has a unique Nash equilibrium (F, F) with the payoff vector $(1/e, 1/e)$ where

$$F(x) = \begin{cases} -\ln(1 - x) & \text{for } x \in [0, 1 - 1/e]\,, \\ 1 & \text{for } x \geq 1 - 1/e\,. \end{cases}$$

Problems

Prove that the non-zero sum noisy duel has no Nash equilibrium. Of course this fact can be easy proved assuming some kind of dibfferentiability and monotonicy of optimal strategies.

Prove that Theorem 5.3.1 gives all Nash equilibria for non-zero sum silent-noisy duel. Of course this fact can be easy proved assuming some kind of differentiability and monotonicy of optimal strategies.

6 Parlour Games

6.1 Cover-up Game

Baston and Bostock [11] considered the following zero-sum game. There are two players, say 1 and 2. Two numbers are chosen from [0, 1] independently by means of uniform distributions. Player 1 looks at the numbers privately, chooses one of the two and opens it to his opponent. Player 2 then accepts either the opened number or the covered (unopened) one, and receives from player 1 his accepted number. Baston and Bostock showed that the optimal strategy of player 1 is to choose the nearest to 1/2 number and open it, the optimal strategy of player 2 is to accept the opened number if and only if it is at least 1/2. The value of the game is 7/12 which is greater than 1/2.

Following Garnaev [48] we consider a generalization of the Baston and Bostock game where player 2 can get some additional information of these two numbers. Namely, having got the opened number, player 2 asks his opponent "Is the opened number greater then the closed one?"There is a positive probability of obtaining a wrong answer to this query. That is, player 1 gives him information according to the following "likehood matrix":

$$\begin{array}{r} \\ \text{opened number} \le \text{closed number} \\ \text{opened number} > \text{closed number} \end{array} \begin{array}{c} \begin{array}{cc} \text{Yes} & \text{No} \end{array} \\ \begin{pmatrix} \bar{\alpha} & \alpha \\ \alpha & \bar{\alpha} \end{pmatrix} \end{array}$$

where $\alpha \in [1/2, 1]$, $\bar{\alpha} = 1 - \alpha$.

The case $\alpha = 1/2$ is equivalent the one where player 2 does not get any additional information about his opponent's numbers. If $\alpha = 1$ then the player receives complete information of relation between numbers.

Let player 1 get two number x and y chosen from $[0, 1]$ independently according to uniform distribution. Introduce the following notations:
$\varphi(x, y)$ is the probability that player 1 opens the minimal of the two numbers,
$\psi_Y(x)$ is the probability that player 2 will accept the opened number x, if player's 1 answer to his query is "Yes,"
$\psi_N(x)$ is the probability that player 2 will accept the opened number x, if player's 1 answer to his query is "No."

It is clear that $\varphi(x,y)$ can be considered as a strategy of player 1 if he has received two numbers x and y, and

$$\varphi(x,y) = \varphi(y,x). \tag{6.1}$$

That is, the order in which player 1 looks through two numbers is indifferent.

A pair $(\psi_{\mathrm{Y}}(x), \psi_{\mathrm{N}}(x))$ can be considered as a strategy of player 2 if he observes number x. Then, the payoff to player 1 is given by

$$\begin{aligned} M(\varphi,(\psi_{\mathrm{Y}},\psi_{\mathrm{N}})) = &\iint\limits_{x\le y} \Big(K(x,y,\varphi,\psi_{\mathrm{Y}},\bar{\alpha}) + K(x,y,\varphi,\psi_{\mathrm{N}},\alpha)\Big)\,dx\,dy \\ &+ \iint\limits_{x>y} \Big(K(y,x,\varphi,\psi_{\mathrm{Y}},\bar{\alpha}) + K(y,x,\varphi,\psi_{\mathrm{N}},\alpha)\Big)\,dx\,dy, \end{aligned}$$

where

$$K(x,y,\varphi,\psi,\alpha) = \alpha\varphi(x,y)\Big(x\psi(x) + y\overline{\psi(x)}\Big) + \bar{\alpha}\overline{\varphi(x,y)}\Big(y\psi(y) + x\overline{\psi(y)}\Big).$$

Clearly, by (6.1),

$$M(\varphi,(\psi_{\mathrm{Y}},\psi_{\mathrm{N}})) = 2\iint\limits_{x\le y} \Big(K(x,y,\varphi,\psi_{\mathrm{Y}},\bar{\alpha}) + K(x,y,\varphi,\psi_{\mathrm{N}},\alpha)\Big)\,dx\,dy.$$

Theorem 6.1.1 *The cover-up game has the value $v(\alpha)$. The optimal strategies of the players are φ^* and $(\psi_{\mathrm{Y}}^*, \psi_{\mathrm{N}}^*)$, where*

$$v(\alpha) = (2/3)(1 - \alpha\delta^3 - \bar{\alpha}\bar{\delta}^3) + (31/24)\bar{\alpha}(1/2 - \delta)^3,$$

$$\varphi^*(x,y) = \begin{cases} 1 & \text{if } x + y \ge 1, \\ 0 & \text{otherwise}, \end{cases}$$

$$\psi_{\mathrm{Y}}^*(x) = \chi_{[\delta,1]}(x), \quad \psi_{\mathrm{N}}^*(x) = \chi_{[\bar{\delta},1]}(x)$$

and

$$\delta = \delta(\alpha) = 2\bar{\alpha}/(1 + 2\bar{\alpha}).$$

Proof Let Player 2 employ the strategy $(\psi_{\mathrm{Y}}^*, \psi_{\mathrm{N}}^*)$, then

$$M(\varphi,(\psi_{\mathrm{Y}}^*,\psi_{\mathrm{N}}^*)) = 2\iint\limits_{x\le y} T(x,y)\varphi(x,y)\,dx\,dy + \text{an expression independing on } \varphi,$$

where

$$T(x,y) = (y - x)\left(1 - \bar{\alpha}(\psi_{\mathrm{Y}}^*(x) + \psi_{\mathrm{N}}^*(y)) - \alpha(\psi_{\mathrm{Y}}^*(y) + \psi_{\mathrm{N}}^*(x))\right).$$

Since player 1 is to minimize $M(\varphi,(\psi^*_{\rm Y},\psi^*_{\rm N}))$ then the optimal strategy φ^* has to be such that for $x \le y$

$$\varphi^*(x,y) = \begin{cases} 1 & \text{if } T(x,y) > 0\,, \\ 0 & \text{if } T(x,y) < 0\,, \\ \text{arbitrary} & \text{if } T(x,y) = 0\,. \end{cases} \tag{6.2}$$

For $\delta < 1/2$ and $\delta + \bar{\delta} = 1$, it is clear that φ^* satisfies (6.2). Thus,

$$M(\varphi,(\psi^*_{\rm Y},\psi^*_{\rm N})) \ge M(\varphi^*,(\psi^*_{\rm Y},\psi^*_{\rm N})) \quad \text{for any strategy} \quad \varphi\,.$$

Let Player 1 employ the strategy φ^*, then

$$M(\varphi^*,(\psi_{\rm Y},\psi_{\rm N})) = 2\int_0^1 L(\varphi^*,\alpha,y)\psi_{\rm Y}(y)\,dy \;+ 2\int_0^1 L(\varphi^*,\bar{\alpha},y)\psi_{\rm N}(y)\,dy$$
$$+ \text{ an expression independing on } \psi_{\rm Y} \text{ and } \psi_{\rm N}\,,$$

where

$$L(\varphi^*,\alpha,y) = \alpha\int_0^y \overline{\varphi^*(x,y)}(y-x)\,dx + \bar{\alpha}\int_y^1 \varphi^*(x,y)(y-x)\,dx\,.$$

It is easy to see that

$$L(\varphi^*,\alpha,y) = \begin{cases} ((1+2\bar{\alpha})y - 2\bar{\alpha})y/2 & \text{if } y \le 1/2\,, \\ ((1+2\alpha)y-1)(1-y)/2 & \text{otherwise}\,. \end{cases}$$

Since

$$L(\varphi^*,\alpha,y)\begin{cases} <0 & \text{if } y < \delta\,, \\ >0 & \text{if } y > \delta \end{cases}$$

and

$$L(\varphi^*,\bar{\alpha},y)\begin{cases} <0 & \text{if } y < \bar{\delta}\,, \\ >0 & \text{if } y > \bar{\delta}\,. \end{cases}$$

It is clear that $M(\varphi^*,(\psi_{\rm Y},\psi_{\rm N}))$ reaches the maxinum on $(\psi^*_{\rm Y},\psi^*_{\rm N})$. Thus,

$$M(\varphi^*,(\psi_{\rm Y},\psi_{\rm N})) \le M(\varphi^*,(\psi^*_{\rm Y},\psi^*_{\rm N})) \quad \text{for any strategy} \quad (\psi_{\rm Y},\psi_{\rm N})\,.$$

Straitforward computation show that

$$M(\varphi^*,(\psi^*_{\rm Y},\psi^*_{\rm N})) = v(\alpha)\,.$$

Now the result follows.

It is interesting to note that the optimal strategy of player 1 is independent of α. The value of the game $v(\alpha)$ varies monotonically from $7/12$ for $\alpha = 1/2$ to $2/3$ for $\alpha = 1$. For $\alpha = 1$ the value of the game is given by

$$2\iint\limits_{x \le y} y\,dy\,.$$

That is the expected maximum of two numbers drawn independently from $[0,1]$ by means of uniform distribution.

6.2 Exchange Games

Consider the following zero-sum game. Two players, say 1 and 2, draw a number x and y, respectively, independently according to a uniform distribution on [0, 1]. After observing his number each player can then choose either to keep his number or to exchange it for the other player's number. If only one of the player chooses exchange it, the exchange takes place with probability p where $p \in (0, 1)$. For convenience we exlude the values $p = 0$ and $p = 1$. The players' 1 payoff is 1 if the number received by him is greater than the number of his opponet. If they receive the same number their payoffs are zero, otherwise the players' 1 payoff is -1. Then, the game can be descibed by the following matrix

$$\begin{array}{cc} & \begin{array}{cc}\text{Exchange} & \text{Refuse}\end{array} \\ \begin{array}{c}\text{Exchange}\\ \text{Refuse}\end{array} & \begin{pmatrix} \operatorname{sgn}(y-x) & p\operatorname{sgn}(y-x)+\bar{p}\operatorname{sgn}(x-y) \\ p\operatorname{sgn}(y-x)+\bar{p}\operatorname{sgn}(x-y) & \operatorname{sgn}(x-y) \end{pmatrix}\end{array}$$

where

$$\operatorname{sgn}(\xi) = \begin{cases} 1 & \text{if } \xi > 0\,, \\ 0 & \text{if } \xi = 0\,, \\ -1 & \text{if } \xi < 0\,. \end{cases}$$

Denote by $\varphi(x)$ the probability that player 1 offers the exchange if he has drawn x, and denote by $\psi(y)$ the probability that player 2 offers the exchange if he has drawn y. It is clear that φ and ϕ can be considered as strategies of the players. Then, the payoff to player 1 is given by

$$M(\varphi, \psi) = \int_0^1\!\!\int_0^1 H(\varphi(x), \psi(y), x, y)\, dx\, dy\,,$$

where

$$\begin{aligned} H(\varphi(x), \psi(y), x, y) &= \operatorname{sgn}(y-x)\varphi(x)\psi(y) \\ &\quad + \left(\overline{\varphi(x)}\psi(y) + \varphi(x)\overline{\psi(y)}\right)(p\operatorname{sgn}(y-x) + \bar{p}\operatorname{sgn}(x-y)) \\ &\quad + \operatorname{sgn}(x-y)\overline{\varphi(x)}\;\overline{\psi(y)}\,. \end{aligned}$$

Thus,

$$\begin{aligned} H(\varphi(x), \psi(y), x, y) &= \big(-1 + 2(p(\varphi(x) + \psi(y)) \\ &\quad + (1-2p)\varphi(x)\psi(y))\big)\operatorname{sgn}(y-x)\,. \end{aligned}$$

Theorem 6.2.1 *The exchange game has value zero. The optimal strategies of the players are* φ^* *and* ψ^*, *where*

$$\varphi^*(x) = \psi^*(x) = \chi_{[0,p]}(x)\,.$$

Proof Let player 2 employ the strategy $\psi^*(x) = \chi_{[0,p]}(x)$, then for any strategy φ of player 1

$$M(\varphi, \psi^*) = 2 \int_0^1 L(x, \psi^*)\varphi(x)\, dx$$
$$+ \text{ an expression independing on } \varphi,$$

where

$$L(x, \psi^*) = p(1 - 2x) + (1 - 2p) \left(\int_x^1 \psi^*(y)\, dy - \int_0^x \psi^*(y)\, dy \right) .$$

It is clear that

$$L(x, \psi^*) = p(1 - 2x) + (1 - 2p) \times \begin{cases} p - 2x & \text{if } x \le p, \\ -p & \text{otherwise}. \end{cases}$$

So, $L(x, \psi^*)$ is strictly decreasing and continuous for $x \in [0, 1]$ such that $L(p, \psi^*) = 0$. Thus,

$$M(\varphi, \psi^*) \le M(\varphi^*, \psi^*) \quad \text{for any strategy} \quad \varphi .$$

By simmetry,

$$M(\varphi^*, \psi) \ge M(\varphi^*, \psi^*) \quad \text{for any strategy} \quad \psi .$$

Now the result follows.

Consider a two-stage variant of the exchange game where players act not simultaneously but consecutively. Namely, first player 1 acts and then player 2. On the first stage of the game player 1 after observing his number either refuses to offer an exchange and then the game stops, or offers to exchange. In the later case player 2 in his turn after observing his number either offer to exchange or refuses to do it. Then, the game can be descibed by the following matrix

$$\begin{array}{l} \\ \text{Exchange} \\ \text{Refuse} \end{array} \begin{array}{c} \begin{array}{cc} \text{Exchange} & \text{Refuse} \end{array} \\ \begin{pmatrix} \operatorname{sgn}(y - x) & p \operatorname{sgn}(y - x) + \bar{p} \operatorname{sgn}(x - y) \\ \operatorname{sgn}(x - y) & \operatorname{sgn}(x - y) \end{pmatrix} \end{array}$$

Thus, the payoff to player 1 is given by

$$M(\varphi, \psi) = \int_0^1 \int_0^1 H(\varphi(x), \psi(y), x, y)\, dx\, dy ,$$

where

$$\begin{aligned}H(\varphi(x),\psi(y),x,y) &= \mathrm{sgn}(y-x)\varphi(x)\psi(y)\\ &\quad+\varphi(x)\overline{\psi(y)}\left(p\,\mathrm{sgn}(y-x)+\bar{p}\,\mathrm{sgn}(x-y)\right)\\ &\quad+\mathrm{sgn}(x-y)\overline{\varphi(x)}\,.\end{aligned}$$

So,

$$H(\varphi(x),\psi(y),x,y) = \left(-1+2\left(p+\bar{p}\psi(y)\right)\varphi(x)\right)\mathrm{sgn}(y-x)\,.$$

Theorem 6.2.2 *The two-stage exchange game has value pc, where* $c = 2p/(1+3p)$. *The optimal strategies of the players are* $\varphi^*(x) = \chi_{[0,c]}(x)$ *and* $\psi^*(x) = \chi_{[0,c/2]}(x)$.

Proof Let player 1 employ the strategy $\varphi^*(x) = \chi_{[0,c]}(x)$, then for any strategy ψ of player 2

$$M(\varphi,\psi^*) = 2(1-p)\int_0^1 T(x,\varphi^*)\psi(x)\,dx$$
$$+ \text{ an expression independing on } \psi\,,$$

where

$$T(x,\varphi^*) = \int_0^x \varphi^*(y)\,dy - \int_x^1 \varphi^*(y)\,dy\,.$$

Thus,

$$T(x,\varphi^*) = \begin{cases} 2x-c & \text{if } x \le c/2\,,\\ c & \text{otherwise}\,.\end{cases}$$

So,

$$T(x,\varphi^*)\begin{cases} <0 & \text{if } x < c/2\,,\\ >0 & \text{if } x > c/2\,.\end{cases}$$

It implies that

$$M(\varphi^*,\psi) \ge M(\varphi^*,\psi^*) \quad \text{for any strategy} \quad \psi\,.$$

Let player 2 employ the strategy $\psi^*(x) = \chi_{[0,p]}(x)$, then for any strategy φ of player 1

$$M(\varphi,\psi^*) = 2\int_0^1 L(x,\psi^*)\varphi(x)\,dx$$
$$+ \text{ an expression independing on } \varphi\,,$$

where

$$L(x,\psi^*) = p(1-2x) + (1-p)\left(\int_x^1 \psi^*(y)\,dy - \int_0^x \psi^*(y)\,dy\right)\,.$$

Thus,

$$L(x,\psi^*) = p(1-2x) + (1-p) \times \begin{cases} c/2 - 2x & \text{if } x \le c/2\,, \\ -c/2 & \text{otherwise}\,. \end{cases}$$

It is clear that $L(x,\psi^*)$ is strictly decreasing and continuous for $x \in [0,1]$ such that $L(c,\psi^*) = 0$ where $c = 2p/(1+3p)$. Thus,

$$M(\varphi,\psi^*) \le M(\varphi^*,\psi^*) \quad \text{for any strategy} \quad \varphi\,.$$

Now the result follows.

Consider another two-stage setup of the exchange game where first player 1 after observing his number either refuses to offer an exchange, or offers to exchange and, then claims his decision to his opponet. Then, player 2 in his turn after observing his number either offer to exchange or refuses to do it. Let $\psi_R(y)$ be the probability that player 2 offers the exchange if he has drawn y and player 1 has refused to exchange, and $\psi_E(y)$ is the probability that player 2 offers the exchange if he has drawn y and player 1 has claimed the exchange. Then, $(\psi_{\mathrm{R}}(y), \psi_{\mathrm{E}}(y))$ can be considered as a strategy of player 2. So, the payoff to player 1 is given by

$$M(\varphi,(\psi_{\mathrm{R}},\psi_{\mathrm{E}})) = \int_0^1\int_0^1 H(\varphi(x),\psi_{\mathrm{R}}(y),\psi_{\mathrm{E}}(y),x,y)\,dx\,dy\,,$$

where

$$\begin{aligned} H(\varphi(x),\psi_{\mathrm{R}}(y),\psi_{\mathrm{E}}(y),x,y) = \operatorname{sgn}(x-y)\Big(&\overline{\varphi(x)}\,\overline{\psi_{\mathrm{R}}(y)} - \varphi(x)\psi_{\mathrm{E}}(y) \\ &+ (1-2p)\big(\overline{\varphi(x)}\psi_{\mathrm{R}}(y) + \varphi(x)\overline{\psi_{\mathrm{E}}(y)}\big)\Big)\,. \end{aligned}$$

Theorem 6.2.3 *The two-stage exchange game has the value $-p\bar{p}/2$. The strategies φ^* and $(\psi^*_{\mathrm{R}},\psi^*_{\mathrm{E}})$ of the players are optimal, where*

$$\varphi^*(x) = \begin{cases} 1 & \text{if } x \in [0,a)\,, \\ \alpha(x) & \text{if } x \in [a,b]\,, \\ 0 & \text{if } x \in (b,1]\,, \end{cases}$$

$$\psi^*_{\mathrm{R}}(x) = \chi_{[0,\mathrm{a}]}(x), \quad \psi^*_{\mathrm{E}}(x) = \chi_{[0,\mathrm{b}]}(x)\,,$$
$$a = p/2, \quad b = (1+p)/2$$

and $\alpha(x) \in [0,1]$ for $x \in [a,b]$ is such that

$$\int_a^b \alpha(x)\,dx = p/2\,.$$

Proof Let player 1 employ the strategy $\varphi^*(x)$ then for any strategy $(\psi_{\mathrm{R}}, \psi_{\mathrm{E}})$ of player 2 we have that

$$M(\varphi^*, (\psi_{\mathrm{R}}, \psi_{\mathrm{E}})) = 2p \int_0^1 T(x, \overline{\varphi^*})\psi_{\mathrm{R}}(x)\, dx + 2\bar{p} \int_0^1 T(x, \varphi^*)\psi_{\mathrm{E}}(x)\, dx$$
$$+ \text{ an expression independing on } \psi_R \text{ and } \psi_{\mathrm{E}},$$

where

$$T(x, \varphi^*) = \int_0^x \varphi^*(y)\, dy - \int_x^1 \varphi^*(y)\, dy.$$

Thus,

$$T(x, \varphi^*) = \begin{cases} 2x - p & \text{if } x \le a, \\ a + \left(\int_a^x - \int_x^b\right) \alpha(y)\, dy & \text{if } x \in (a, b), \\ p & \text{otherwise} \end{cases}$$

and

$$T(x, \overline{\varphi^*}) = \begin{cases} -\bar{p} & \text{if } x \le a, \\ 2x - a - 1 - \left(\int_a^x - \int_x^b\right) \alpha(y)\, dy & \text{if } x \in (a, b), \\ 2x - \bar{p} & \text{otherwise}. \end{cases}$$

Thus, $T(x, \varphi^*)$ and $T(x, \bar{\varphi}^*)$ are continuous and nondecreasing such that $T(a, \varphi^*) = 0$ and $T(b, \bar{\varphi}^*) = 0$. So,

$$M(\varphi^*, (\psi_{\mathrm{R}}, \psi_{\mathrm{E}})) \ge M(\varphi^*, (\psi^*_{\mathrm{R}}, \psi^*_{\mathrm{E}})) \quad \text{for any strategy} \quad (\psi_{\mathrm{R}}, \psi_{\mathrm{E}}).$$

Let player 2 employ the strategy $(\psi^*_{\mathrm{R}}, \psi^*{}_{rmE})$ then for any strategy φ of player 1

$$M(\varphi, (\psi^*_{\mathrm{R}}, \psi^*_{\mathrm{E}})) = -2p \int_0^1 L(x, \psi^*_{\mathrm{R}}, \psi^*_{\mathrm{E}})\varphi(x)\, dx$$
$$+ \text{ an expression independing on } \varphi,$$

where

$$L(x, \psi^*_{\mathrm{R}}, \psi^*_{\mathrm{E}}) = \int_0^x \left(p\overline{\psi^*_{\mathrm{R}}(y)} + \bar{p}\psi^*_{\mathrm{E}}(y)\right) dy - \int_x^1 \left(p\overline{\psi^*_{\mathrm{R}}(y)} + \bar{p}\psi^*_{\mathrm{E}}(y)\right) dy.$$

Since $\psi^*_{\mathrm{R}}(x) = \chi_{[0,a]}(x)$ and $\psi^*_{\mathrm{E}}(x) = \chi_{[0,b]}(x)$, then

$$L(x, \psi^*_{\mathrm{R}}, \psi^*_{\mathrm{E}}) = \begin{cases} 2\bar{p}x - (\bar{p}a + p\bar{b}) & \text{if } x \le a, \\ 0 & \text{if } x \in (a, b), \\ 2px + \bar{p}a - p(1 + b) & \text{otherwise}. \end{cases}$$

Hence,

$$L(x, \psi_{\mathrm{R}}^*, \psi_{\mathrm{E}}^*) \begin{cases} < 0 & \text{if } x \le a\,, \\ = 0 & \text{if } x \in (a, b)\,, \\ > 0 & \text{otherwise}\,. \end{cases}$$

Thus, it implies that

$$M(\varphi, (\psi_{\mathrm{R}}^*, \psi_{\mathrm{E}}^*)) \le M(\varphi^*, (\psi_{\mathrm{R}}^*, \psi_{\mathrm{E}}^*)) \quad \text{for any strategy} \quad \varphi\,.$$

The result now follows.

6.3 Poker Games

Investigation of exchange games was initiated by Brams, Kilgour and Davis [24]. Garnaev [45] and Sakaguchi [98] studied a different scenarios of the game. Note that by its structure the exchange games are similar to poker games (Karlin [57], Sakaguchi [97]). Consider four examples of poker game.

6.3.1 First Set-up of Poker Game

Each of two players, say 1 and 2, receives a hand x and y respectively in $[0, 1]$ according to a uniform distribution and chooses one of two alternatives: Fold or Bet the amount $1 + A$ where $A > 0$. Here 1 is the ante to the game and A is a given positive number. If one player bets and the other folds, then the player who made the bet wins the pot. If both players fold the game is a draw and no payoffs return. If both players bet then the players compare hands and the player with the higher hand wins the pot. Then, the game can be descibed by the following matrix

$$\begin{array}{cc} & \begin{array}{cc} \text{Fold} & \text{Bet} \end{array} \\ \begin{array}{c} \text{Fold} \\ \text{Bet} \end{array} & \begin{pmatrix} 0 & -1 \\ 1 & (1+A)\,\mathrm{sgn}(x-y) \end{pmatrix} \end{array}$$

Theorem 6.3.1 *By symmetry, the value of the game is zero. The player has the same optimal strategies*

$$\varphi^*(x) = \psi^*(x) = \chi_{[b,1]}(x)\,,$$

where $b = A/(1 + A)$, and $\varphi(x)$ ($\psi(y)$) is the probability that player 1 (2) bets on the hand x (y).

Proof It is clear that

$$M(\varphi, \psi) = \int_0^1 L(x, \psi)\varphi(x)\,dx + \text{an expression not depending on } \varphi\,,$$

where

$$L(x,\psi) = 1 + (1+A)\left(\int_0^x \psi(y)\,dy - \int_x^1 \psi(y)\,dy\right).$$

Thus,

$$L(x,\chi_{[b,1]}) = \begin{cases} 1+(1+A)(2x-b-1) & \text{if } x > b, \\ -A+(1+A)b & \text{if } x \le b. \end{cases}$$

Hence,

$$L(x,\chi_{[b,1]}) \begin{cases} = 0 & \text{for } x \le b, \\ > 0 & \text{for } x > b. \end{cases}$$

This implies that

$$M(\varphi,\psi^*) \le M(\varphi^*,\psi^*) \le M(\varphi^*,\psi) \quad \text{for any strategies} \quad \varphi \quad \text{and} \quad \psi.$$

6.3.2 Second Set-up of Poker Game

Each of two players, say 1 and 2, receives a hand x and y respectively in $[0,1]$ according to a uniform distribution and chooses one of two alternatives: Fold or Bet the amount $1+A$ where $A>0$. First player 1 acts. He may either bet A more units or fold and forfeit his initial ante. If player 1 bets, then player 2 has two choices: he may either fold, losing his initial ante, or bet A units and see the player's 1 hand. If player 2 sees, the two players compare hands and and the one with the highter hand wins the pot. Then, the game can be descibed by the following matrix

$$\begin{array}{c} \\ \text{Fold} \\ \text{Bet} \end{array} \begin{array}{c} \begin{array}{cc} \text{Fold} & \text{Bet} \end{array} \\ \begin{pmatrix} -1 & -1 \\ 1 & (1+A)\operatorname{sgn}(x-y) \end{pmatrix} \end{array}$$

Theorem 6.3.2 *The value of the game is $-c^2$, where $c = A/(A+2)$. The strategies $\varphi^*(x)$ and $\psi^*(y)$ of the players are optimal where*

$$\varphi^*(x) = \begin{cases} \alpha(x) & \text{if } x \in [0,c], \\ 1 & \text{if } x \in (c,1], \end{cases}$$

$$\psi^*(y) = \chi_{[c,1]}(y),$$

and $\alpha : [0,c] \to [0,1]$ is a function such that

$$\int_0^c \alpha(x)\,dx = c\bar{c}.$$

Proof It is clear that

$$M(\varphi,\psi)=\int_0^1 L(x,\psi)\varphi(x)\,dx+\text{an expression not depending on }\varphi$$

and

$$M(\varphi,\psi)=\int_0^1 T(y,\varphi)\psi(y)\,dy+\text{an expression not depending on }\psi,$$

where

$$L(x,\psi)=2-\int_0^1\psi(y)\,dy+(1+A)\left(\int_0^x\psi(y)\,dy-\int_x^1\psi(y)\,dy\right)$$

and

$$T(y,\varphi)=-\int_0^1\varphi(x)\,dx-(1+A)\left(\int_0^y\varphi(x)\,dx-\int_y^1\varphi(x)\,dx\right).$$

Thus,

$$L(x,\chi_{[c,1]})=\begin{cases}1+c-(1+A)(1-c) & \text{if } x\le c\\ 1+c+(1+A)(2x-c-1) & \text{if } x>c.\end{cases}$$

This implies that

$$L(x,\chi_{[c,1]})\begin{cases}=0 & \text{for } x\le c,\\ >0 & \text{for } x>c.\end{cases}$$

So,

$$M(\varphi,\psi^*)\le M(\varphi^*,\psi^*)\quad\text{for any strategy}\quad\psi.$$

Also,

$$T(y,\varphi_*)=c-1-\int_0^c\alpha(x)\,dx$$
$$+(1+A)\times\begin{cases}\left(1-c+\int_y^c\alpha(x)\,dx-\int_0^y\alpha(x)\,dx\right) & \text{for } y\le c,\\ \left(1+c-2y-\int_0^c\alpha(x)\,dx\right) & \text{for } y>c.\end{cases}$$

Since $\int_y^c\alpha(x)\,dx-\int_0^y\alpha(x)\,dx$ is non-increasing for $y\in[0,c]$ and $\int_0^c\alpha(x)\,dx=c\bar{c}$, we have that

$$T(y,\varphi_*)\begin{cases}\geq 0 & \text{for } y \leq c\,,\\ = 0 & \text{for } y = c\,,\\ < 0 & \text{for } y > c\,.\end{cases}$$

This implies that

$$M(\varphi^*,\psi^*) \leq M(\varphi^*,\psi) \quad \text{for any strategy} \quad \psi\,.$$

The result now follows.

6.3.3 Third Set-up of Poker Game

Following Sakaguchi and Mazalov [100] suppose that in the one-stage poker in the case Bet-Bet players 1 and 2 draw another card, say u and v respectively, and use the card with the greater value throwing away the card with the smaller value. Thus, they compare $\max\{x,u\}$ and $\max\{y,v\}$. Then, the game can be described as follows

$$\begin{array}{cc} & \begin{array}{cc}\text{Fold} & \text{Bet}\end{array} \\ \begin{array}{c}\text{Fold}\\ \text{Bet}\end{array} & \begin{pmatrix} 0 & -1 \\ 1 & (1+A)\operatorname{sgn}(\max\{x,u\}-\max\{y,v\}) \end{pmatrix}\end{array}$$

Theorem 6.3.3 *By symmetry the value of the game is zero. The players have the common optimal strategies and*

$$\varphi^*(x) = \psi^*(y) = \chi_{[c,1]}(y)\,,$$

where c is $\max\{0,(1-3/(1+A))^{1/3}\}$.

Proof It is clear that

$$M(\varphi,\psi) = \int_0^1 L(x,\psi)\varphi(x)\,dx + \text{an expression not depending on } \varphi\,,$$

where

$$L(x,\psi) = 1 + (1+A)\left(\int_0^x x^2\psi(y)\,dy - \int_x^1 y^2\psi(y)\,dy\right)\,.$$

Thus,

$$L(x,\chi_{[c,1]}) = \begin{cases} 1+(1+A)(c^3-1)/3 & \text{if } x \leq c\,,\\ 1-(1+A)/3-(1+A)cx^2+4(1+A)x^3/3 & \text{if } x > c\,.\end{cases}$$

Hence,

$$L(x,\chi_{[b,1]})\begin{cases} = 0 & \text{for } x \leq c\,,\\ > 0 & \text{for } x > c\,.\end{cases}$$

This implies that

$$M(\varphi,\psi^*) \leq M(\varphi^*,\psi^*) \leq M(\varphi^*,\psi) \quad \text{for any strategies} \quad \varphi \quad \text{and} \quad \psi.$$

The result now follows.

6.3.4 Forth Set-up of Poker Game

Suppose that in cases Fold-Bet, Bet-Fold and Fold-Fold of the one-stage poker the players who has chosen Bet draw another card. Say, for player 1 and 2 it might be u and v respectively. The player who drawn a new card uses the card with the greater value throwing away the card with the smaller value. After that they compare their hands. For Fold-Bet, Bet-Fold the prize of the winner is 1, for Bet-Bet it is $1+A$. Then, the game can be described as follows

$$\begin{array}{c} \\ \text{Fold} \\ \text{Bet} \end{array} \begin{array}{c} \begin{array}{cc} \text{Fold} & \qquad\qquad\qquad \text{Bet} \end{array} \\ \begin{pmatrix} 0 & \operatorname{sgn}(x - \max\{y, v\}) \\ \operatorname{sgn}(\max\{x, u\} - y) & (1 + A)\operatorname{sgn}(\max\{x, u\} - \max\{y, v\}) \end{pmatrix} \end{array}$$

Theorem 6.3.4 *The value of the game is zero. The players have the common optimal strategies and*

$$\varphi^*(x) = \psi^*(y) = \chi_{[c,1]}(y)\,,$$

where c is the unique root of the equation

$$-1 - (1 + A)/3 + 2c + (1 + A)c^3/3 = 0\,.$$

Proof It is clear that

$$M(\varphi, \psi) = \int_0^1 L(x, \psi)\varphi(x)\,dx + \text{an expression not depending on } \varphi$$

and

$$M(\varphi, \psi) = -\int_0^1 L(y, \varphi)\psi(y)\,dy + \text{an expression not depending on } \psi\,,$$

where

$$L(x, \psi) = x^2 - 2\int_0^1 \psi(y)\,dy + 2\int_x^1 y\psi(y)\,dy$$
$$+ 2\int_0^x x\psi(y)\,dy + (1 + A)\left(\int_0^x x^2\psi(y)\,dy - \int_x^1 y^2\psi(y)\,dy\right).$$

Thus,

$$L(x, \chi_{[c,1]}) = 2c - 1 - c^2 - (1 + A)/3 + \begin{cases} (1 + A)c^3/3 + x^2 & \text{if } x \le c\,, \\ (2 - (1 + A)c)x^2 - 2cx + 4(1 + A)x^3/3 & \text{if } x > c\,. \end{cases}$$

Hence,

$$L(x, \chi_{[b,1]}) \begin{cases} < 0 & \text{for } x \leq c\,, \\ = 0 & \text{for } x = c\,, \\ > 0 & \text{for } x > c\,. \end{cases}$$

This implies that

$$M(\varphi, \psi^*) \leq M(\varphi^*, \psi^*) \leq M(\varphi^*, \psi) \quad \text{for any strategies} \quad \varphi \quad \text{and} \quad \psi\,.$$

The result now follows.

6.4 One-Person Games with a Threshold

Consider the following one person game denoted by Γ_1. A player receives a hand, say x, drawn from [0, 1] according to uniform distribution. He can accept this hand, or request another one. If he accepts it, the game is over and his payoff is 1. If he requests another hand, he receives it, say y, drawn independently from [0,1] according to uniform distribution. In this case if $x + y \leq 1$, his payoff is A, where $A > 1$ and 0 otherwise. The player is to maximize the payoff. In this game the player has an alternative either to keep a reliable but small prize, or to risk to lose it in hope to get a great one. A strategy of the player can be described as $\varphi(x)$ where $\varphi(x)$ is the probability that the player has chosen to request the next hand, after observing the hand x received on the first stage of the game. Then, the payoff $M(\varphi)$ to the player is given by

$$M(\varphi) = \int_0^1 \overline{\varphi(x)}\, dx + A \int_0^1 \varphi(x)(1 - x)\, dx\,,$$

where $\bar{\xi} = 1 - \xi$.

Theorem 6.4.1 *The optimal strategy φ^* of the player and his payoff v in the game Γ_1 are given as follows*

$$\varphi^*(x) = \chi_{[0,1-1/A]}(x)\,,$$
$$v = A/2 + 1/(2A)\,,$$

where χ_E is the indicator function of a set E, thus, $\chi_E(x) = 1$ if $x \in E$ and $\chi_E(x) = 0$ otherwise.

Proof It is clear that

$$M(\varphi) = 1 + \int_0^1 (A - 1 - Ax)\, \varphi(x)\, dx\,.$$

The result now follows.

Theorem 6.4.1 implies that $\lim_{A\to\infty} v/A = 1/2$.

Consider a generalization of the game Γ_1 denoted by Γ_1' where on the second stage of the game the player can request one or two additional cards and his payoff for i requested cards is A_i, where $i = 1, 2$, if the sum of all received him cards is at most 1 and 0 otherwise. We assume that $1 < A_1 < A_2$. Then, the payoff $M(\varphi)$ to the player is given by

$$M(\varphi) = \int_0^1 \varphi_0(x)\,dx + A_1 \int_0^1 \varphi_1(x)(1-x)\,dx + A_2 \int_0^1 \varphi_2(x)(1-x)^2/2\,dx\,,$$

where $\varphi = (\varphi_0, \varphi_1, \varphi_2)$ and $\varphi_i(x)$ for $i \in [0,2]$ is the probability that the player has chosen to request i cards, after observing the hand x received on the first stage of the game. It is clear that $\varphi_0(x) + \varphi_1(x) + \varphi_2(x) = 1$.

Our next theorem proves that, in general, prizes have different degree of attraction for the player.

Theorem 6.4.2 *The optimal strategy φ^* of the player and the payoff v in the game Γ_1' are given as follows*
(a) if $2A_1 \geq A_2$ then

$$\begin{aligned}\varphi_0^*(x) &= \chi_{(1-1/A_1,1]}(x),\\ \varphi_1^*(x) &= \chi_{[0,1-1/A_1]}(x),\\ \varphi_2^*(x) &= 0,\\ v &= A_1/2 + 1/(2A_1),\end{aligned}$$

(b) if $2A_1 < A_2$ and $A_2 < 2A_1^2$ then

$$\begin{aligned}\varphi_0^*(x) &= \chi_{[1-1/A_1,1]}(x),\\ \varphi_1^*(x) &= \chi_{[1-2A_1/A_2,1-1/A_1)}(x),\\ \varphi_2^*(x) &= \chi_{[0,1-2A_1/A_2)}(x),\\ v &= (A_2^3 + 12A_1A_2 - 8A_1^3)/(6A_2^2),\end{aligned}$$

(c) if $2A_1 < A_2$ and $A_2 \geq 2A_1^2$ then

$$\begin{aligned}\varphi_0^*(x) &= \chi_{(1-\sqrt{2/A_2},1]}(x),\\ \varphi_1^*(x) &= 0,\\ \varphi_2^*(x) &= \chi_{[0,1-\sqrt{2/A_2}]}(x),\\ v &= (4\sqrt{2/A_2} + A_2)/6.\end{aligned}$$

Theorem 6.4.2(c) implies that $\lim_{A_2\to\infty} v/A_2 = 1/6$ for a fixed A_1 Thus, the second card on the second stage of the game Γ_1' weights three times less then the card in Γ_1.

Consider another one person game denoted by Γ_2 being a modification of Γ_1 where the payoff of the player is the hand x on the first stage, and the sum of hands $x+y$, if $x+y \le 1$, and 0 otherwise on the second stage of the game. Then, the payoff $M(\varphi)$ to the player is given by

$$M(\varphi) = \int_0^1 x\overline{\varphi(x)}\,dx + \int_0^1 \varphi(x) \int_0^{1-x} (x+y)\,dy\,dx\,.$$

Theorem 6.4.3 *The optimal strategy φ^* of the player and his payoff v in the game Γ_2 are given as follows*

$$\varphi^*(x) = \chi_{[0,\sqrt{2}-1]}(x) \quad (= \chi_{[0,0.414]}(x))\,,$$
$$v = (2\sqrt{2}-1)/3 \quad (= 0.609)\,.$$

In a generalization of the game Γ_2 where the player on the second stage could choose one or two cards, his optimal strategy is the same as in Γ_2. So, the player never risks to request two cards. This result matches Theorem 19.4.2(a) where the prize for additional risk also is not enough attractive for the player.

6.5 Two-Person Games

Consider the following two-person zero-sum game denoted by Γ_1^1. Each of two players, say 1 and 2, receives a hand x and y respectively in $[0,1]$ according to uniform distribution and chooses one of two alternatives: Open (O) or Next (N). If they both choose Open, the players compare hands and the player with the higher hand wins 1. If a player chooses Open and the other chooses Next, the player, who has chosen Next, wins 1. If both players choose Next then players 1 and 2 receives another hand u and v respectively from $[0,1]$ according to uniform distribution. In this case both players get 0 if $x+u>1$ and $y+v>1$. If $x+u\le 1$ and $y+v\le 1$, the players compare hands and the player with the higher sum of hands wins A. If the sum of hands of a player is at most 1 while the sum of hands of his rival is greater that 1, then the player with the less sum of hands wins A. Thus, the game can be described by the following matrix

$$\begin{array}{cc} & \begin{array}{cc} \mathrm{O} & \quad\mathrm{N} \end{array} \\ \begin{array}{c} \mathrm{O} \\ \mathrm{N} \end{array} & \begin{pmatrix} \operatorname{sgn}(x-y) & -1 \\ 1 & B \end{pmatrix} \end{array}$$

where

$$B = \begin{cases} 0 & \text{if } x+u>1 \text{ and } y+v>1\,, \\ A & \text{if } x+u\le 1 \text{ and } y+v>1\,, \\ -A & \text{if } x+u>1 \text{ and } y+v\le 1\,, \\ A\operatorname{sgn}(x+u-y-v) & \text{if } x+u\le 1 \text{ and } y+v\le 1\,. \end{cases}$$

A strategy of player 1 can be described as $\varphi(x)$, where $\varphi(x)$ is the probability that the player has chosen Next after observing the hand x received on the first stage of the game. Similarly, a strategy of player 2 can be given by the probability $\psi(y)$. Then, the payoff $M(\varphi, \psi)$ to player 1 is given by

$$M(\varphi,\psi) = \int_0^1\int_0^1 \left(\operatorname{sgn}(x-y)\overline{\varphi(x)}\,\overline{\psi(y)} + \varphi(x)\overline{\psi(y)} - \overline{\varphi(x)}\psi(y)\right) dx\,dy$$
$$+ A\int_0^1\int_0^1 \varphi(x)\psi(y)\Big(\int_0^{1-x}\int_0^{1-y} \operatorname{sgn}(x+u-y-v)$$
$$+ \int_0^{1-x}\int_{1-y}^1 1 - \int_{1-x}^1\int_0^{1-y} 1\Big)\, du\,dv\,dx\,dy\,.$$

Theorem 6.5.1 *By symmetry the value of the game Γ_1^1 is zero. The players have the same optimal strategies*

$$\varphi^*(x) = \psi^*(x) = \chi_{[0,c^*]}(x)\,,$$

where $c^* = \min\{c, 1\}$ and c is the unique positive root of the following equation

$$2 - c - Ac^3/2 = 0\,.$$

Proof It is clear that

$$M(\varphi,\psi) = \int_0^1 L(x,\psi)\varphi(x)\,dx + \text{an expression not depending on } \varphi\,,$$

where

$$L(x,\psi) = -\int_0^1 \operatorname{sgn}(x-y)\,dy + \int_0^x \psi(y)\,dy - \int_x^1 \psi(y)\,dy + 1$$
$$+ A\Big(\int_0^1 (y-x)\psi(y)\,dy + \int_0^x (1-x)(x-y)\psi(y)\,dy$$
$$- \int_x^1 (1-y)(y-x)\psi(y)\,dy\Big)\,.$$

Straitforward computation shows that

$$L(x,\chi_{[0,c]}) = \begin{cases} 2+c+(Ac^2/2-2)x - Acx^2 & \text{if } x \ge c \\ 2-c+Ac^3/3 - Ac^2x/2 - Ax^3/3 & \text{if } x < c\,. \end{cases}$$

$L(x, \chi_{[0,c]})$ is continuous in $[0, 1]$ and decreasing in $[0, c]$. Since $L(c, \chi_{[0,c]}) = 0$ and $L'(c + 0, \chi_{[0,c]}) < 0$ then $L(x, \chi_{[0,c]})$ is also decreasing in $[0, c]$. So,

$$L(x, \chi_{[0,c]}) \begin{cases} > 0 & \text{for } x < c, \\ = 0 & \text{for } x = c \\ < 0 & \text{for } x > c. \end{cases}$$

It is clear that $c \leq 1$ for $A \geq 2$ and $c > 1$ for $A < 2$. Thus,

$$M(\varphi, \psi^*) \leq M(\varphi^*, \psi^*) \quad \text{for any strategy} \quad \varphi.$$

The result now follows.

Remark 6.5.1 *For $A < 2$ the players always choose Next but for $A \geq 2$ the players prefer more cautious strategy. This interesting phenomena is caused by the fact that for $A < 2$ the payoffs of the players for O-N, N-O and N-N are commensurable and the players do not make a great difference between them. For $A > 2$ this difference becomes essential for them.*

Consider a variant of the game Γ_1^1 denoted by Γ_2^1 where the players compare their hands in all cases. For O-O, N-O and O-N the prize of the winner is 1 but for N-N this prize is A. Thus, the game can be described as follows

$$\begin{array}{cc} & \begin{array}{cc} \quad O \quad\quad\quad\quad & \quad\quad N \end{array} \\ \begin{array}{c} O \\ \\ N \end{array} & \left(\begin{array}{cc} \text{sgn}(x - y) & \begin{array}{l} 1 \quad\quad\quad\quad \text{if } y + v > 1 \\ \text{sgn}(x - y - v) \ \text{if } y + v \leq 1 \end{array} \\ \begin{array}{l} -1 \quad\quad\quad\quad\quad \text{if } x + u > 1 \\ \text{sgn}(x + u - y) \ \text{if } x + u \leq 1 \end{array} & B \end{array} \right) \end{array}$$

where

$$B = \begin{cases} 0 & \text{if } x + u > 1 \text{ and } y + v > 1, \\ A & \text{if } x + u \leq 1 \text{ and } y + v > 1, \\ -A & \text{if } x + u > 1 \text{ and } y + v \leq 1, \\ A \, \text{sgn}(x + u - y - v) & \text{if } x + u \leq 1 \text{ and } y + v \leq 1. \end{cases}$$

Theorem 6.5.2 *The value of the game Γ_2^1 is zero. The players have the same optimal strategies*

$$\varphi^*(x) = \psi^*(x) = \chi_{[0,c]}(x),$$

where c is the unique positive root of the following equation

$$1 - c - c^2 - Ac^3/2 = 0.$$

Proof It is clear that the payoff $M(\varphi, \psi)$ to player 1 is given by

$$
\begin{aligned}
M(\varphi,\psi) &= \int_0^1\!\!\int_0^1 \operatorname{sgn}(x-y)\overline{\varphi(x)}\,\overline{\psi(y)} \\
&+ \int_0^1\!\!\int_0^1 \varphi(x)\overline{\psi(y)}\Big(\int_0^{1-x} \operatorname{sgn}(x+u-y)\,du - \int_{1-x}^1 1\,du\Big) \\
&+ \int_0^1\!\!\int_0^1 \overline{\varphi(x)}\psi(y)\Big(\int_0^{1-y} \operatorname{sgn}(x-y-v)\,dv + \int_{1-y}^1 1\,dv\Big) \\
&+ A\int_0^1\!\!\int_0^1 \varphi(x)\psi(y)\Big(\int_0^{1-x}\!\!\int_0^{1-y} \operatorname{sgn}(x+u-y-v) \\
&+ \int_0^{1-x}\!\!\int_{1-y}^{1} 1 - \int_{1-x}^{1}\!\!\int_0^{1-y} 1\Big)\,du\,dv\,dx\,dy\,.
\end{aligned}
$$

Thus,

$$
M(\varphi,\psi) = \int_0^1 L(x,\psi)\varphi(x)\,dx + \text{an expression not depending on } \varphi
$$

and

$$
M(\varphi,\psi) = -\int_0^1 L(y,\varphi)\psi(y)\,dy + \text{an expression not depending on } \psi\,,
$$

where

$$
\begin{aligned}
L(x,\psi) &= 1 - 2x - x^2 + \int_0^x \psi(y)\,dy - \int_x^1 \psi(y)\,dy \\
&+ A\Big(\int_0^1 (y-x)\psi(y)\,dy + \int_0^x (1-x)(x-y)\psi(y)\,dy \\
&- \int_x^1 (1-y)(y-x)\psi(y)\,dy\Big)\,.
\end{aligned}
$$

Straitforward computation shows that

$$
L(x,\chi_{[0,c]}) = \begin{cases} 1+c+(Ac^2/2-2)x-(Ac+1)x^2 & \text{if } x \ge c\,, \\ 1-c+Ac^3/3-Ac^2x/2-x^2-Ax^3/3 & \text{if } x < c\,. \end{cases}
$$

Thus,

$$
L(x,\chi_{[0,c]}) \begin{cases} >0 & \text{for } x<c\,, \\ =0 & \text{for } x=c\,, \\ <0 & \text{for } x>c\,. \end{cases}
$$

This implies that

$$M(\varphi,\psi^*) \leq M(\varphi^*,\psi^*) \leq M(\varphi^*,\psi) \quad \text{for any strategies} \quad \varphi \quad \text{and} \quad \psi\,.$$

The result now follows.

It is clear that for a fixed A the switch point in game Γ_2^1 is less than the one in Γ_1^1.

References

1. Alpern, S. (1974): The Search Game with Mobile Hider on the Circle. In: Roxin, E.O., Liu, T.P., Stenberg, R.L. (Eds.): Differential Games and Control Theory. Marcel Dekker, New York, 181-200
2. Alpern, S. (1985): Search for Point in Interval, with High-Low Feedback. Math. Proc. Camb. Phil. Soc. 98, 569-578
3. Alpern, S. (1992): Infiltration Games on Arbitrary Graphs. Journal of Mathematical Analisys and Applications 163, 286-288
4. Alpern, S., Asic, M. (1986): Ambush Strategies in Search Games on Graphs. SIAM Journal on Control and Optimization 24, 66-75
5. Anderson, E.J., Aramendia, M.A., (1990): The Search Game on a Network with Immobile Hider. Networks 20, 817-844
6. Auger, J.M., (1991): An Infiltration Game on k Arcs. Naval Research Logistics 38, 511-529
7. Avenhaus, R., Canty, M., Kilgour, D.M., Stengel, B. and Zamir, S. (1996): Inspection Games in Arms Control. European Journal of Operations Research (to appear)
8. Basar, T., Olsder, G.J. (1982): Dynamic Noncooperative Game Theory. Academic Press, New York
9. Baston, V.J., Bostock, F.A. (1985): A High-Low Search Game on the Unit Interval. Math. Proc. Camb. Phil. Soc. 97, 345-348
10. Baston, V.J., Bostock, F.A. (1987): A Continuous Game of Ambush. Naval Research Logistics 34, 645-654
11. Baston, V.J., Bostock, F.A. (1988): An Evasion Game with Barriers. SIAM Journal on Control and Optimization 26, 1099-1105
12. Baston, V.J., Bostock, F.A. (1988): A Simple Cover-up Game. American Mathematics Monthly 95, 850-854
13. Baston, V.J., Bostock, F.A. (1989): A One-Dimensional Helicopter-Submarine Game. Naval Research Logistics 36, 479-490
14. Baston, V.J., Bostock, F.A. (1991): A Generalized Inspection Game. Naval Research Logistics 38, 171-182
15. Baston, V.J., Bostock, F.A., Ferguson, T.S. (1989): The Number Hides Game. Proceedings of the American Mathematical Society 107, 437-447
16. Baston, V.J., Garnaev, A.Y. (1995): A Teraoka-Type Two-Person Nonzero-sum Silent Duel. Journal of Optimization Theory and Applications 87, 539-551
17. Baston, V.J., Garnaev, A.Y. (1996): A Fast Infiltration Game on n Arcs. Naval Research Logistics 43, 481-489
18. Baston, V.J., Garnaev, A.Y. (1997): A Non-zero-sum War of Attrition. ZOR - Mathematical Methods of Operations Research 45, 197-211
19. Baston, V.J., Garnaev, A.Y. (1998): On a Game in Manufacturing (submitted)
20. Baston, V.J., Garnaev, A.Y. (1998): A Search Game with a Protector. Naval Research Logistics (to appear)

21. Beck, A. (1964): On the Linear Search Problem. Israel Journal of Mathematics 2, 221-228
22. Beck, A., Newman, D.J. (1970): Yet More on the Linear Search Problem. Israel Journal of Mathematics 8, 419-429
23. Bellman, R. (1963): An Optimal Search Problem. SIAM Review 5, 274
24. Brams, S.J., Kilgour, D.M., Davis, M.D. (1993): Unraveling in Games of Sharing and Exchange. In: Binmore, K., Kirman, A., Tani, P. (Eds.): Fronties of Game Theory. MIT Press, Cambridge 194-212
25. Croucher, J.S. (1975): Application of the Fundamental Theorem of Game to an Example Concerning Antiballistic Missle Defence. Naval Research Logistics Quarterly 22, 197-203
26. Danskin, J.M. (1967): The Theory of Max-Min. Springer, Berlin
27. Danskin, J.M. (1986): A Helicopter Versus Submarine Search Game. Operations Research 16, 509-517
28. Dobbie, J.M. (1974): A Two Cell Modell of Search for a Moving Target. Operations Research 22, 79-92
29. Dresher, M. (1962): A Sampling Inspection Problem in Arms Control Agreements: a Game Theoretic Analysis. Memorandum No. RM-2972-ARPA, The RAND Corporation, Santa Monica, California
30. Eagle, J.N., Washburn, A.R. (1991): Cumulative Search-Evasion Games. Naval Research Logistics 38, 495-510
31. Fitzgerald, C.H. (1979) The Princess and Monster Differential Game. SIAM Journal on Contol and Optimization 17, 700-712
32. Ferguson, T.S., Melolidakis, C. (1998): On The Inspection Game. Naval Research Logistics 45, 327-324
33. Friedman, L. (1958): Game-Theory Model in the Allocation of Advertising Expenditures. Operations Research 6, 699-709
34. Fudenberg, D., Tirole, J. (1993): Game Theory. MIT Press, Cambridge, Massachusetts
35. Gal, S. (1972): A General Search Problem. Israel Journal of Mathematics 12, 32-45
36. Gal, S. (1974): Minimax Solutions for Linear Search Problems. SIAM Journal of Applied Mathematics 27, 17-30
37. Gal, S. (1974): A Discrete Search Game. SIAM Journal of Applied Mathematics 27, 641-648
38. Gal, S. (1978): A Stochastic Search Game. SIAM Journal of Applied Mathematics 34, 205-210
39. Gal, S. (1979): Search Games with Mobile and Immobile Hider. SIAM Journal on Control and Optimization 17, 99-122
40. Gal, S. (1980): Search Games. Pergamon Perss, New York
41. Gal, S., Chazan, D. (1976): On the Optimality of the Exponential Functions for Some Minimax Problems. SIAM Journal of Applied Mathematics 30, 324-348
42. Garnaev, A.Y. (1990): A Discrete Game with Time Lag on a Line. Automation and Remote Control 51, 1147-1154
43. Garnaev, A.Y. (1991): A Search Game in a Rectangle. Journal of Optimization Theory and Applications 69, 531-542
44. Garnaev, A.Y. (1992): A Remark on the Princess and Monster Search Game. International Journal of Game Theory 20, 269-276
45. Garnaev, A.Y. (1992): On a Simple MIX Game. International Journal of Game Theory 21, 237-247
46. Garnaev, A.Y. (1993): A Remark on a Helicopter-Submarine Game. Naval Research Logistics 40, 745-753

47. Garnaev, A.Y. (1994): A Remark on a Customs and Smuggler Game. Naval Research Logistics 41, 287-293
48. Garnaev, A.Y. (1995): The Value of Sample Information in a Cover-up Game. Mathematica Japonica 41, 253-259
49. Garnaev, A.Y. (1996): On a Sakaguchi's Problem in Non-zero-sum Games of Timing. Mathematica Japonica 44, 291-301
50. Garnaev, A.Y. (1997): On a Ruckle's Problem in Discrete Games of Ambush. Naval Research Logistics 44, 353-364
51. Garnaev, A.Y. (1999): On a Sakaguchi's Problem in Noisy and Noisy-Silent Games of Timing. Mathematica Japonica 49, 351-361
52. Garnaev, A.Y., Garnaeva, G.Y., Goutal, P. (1997): On the Infiltration Game. International Journal of Game Theory 26, 215-221
53. Hamers, H. (1993): A Silent Duel over a Cake. ZOR-Methods and Models of Operational Research 37, 119-127
54. Hendrics, K., Weiss, A., Wilson, C., (1988): The War of Attrition in Continuous time with Complete Informatiom. International Economic Review 29, 663-680
55. Iida, K., Hohzaki, R., Satō, K. (1994): Hide-and-Search Game with the Risk Criterion. Journal of Operations Research Society of Japan 37, 287-296
56. Isaacs, R. (1973): Some Fundamentals of Differential Games. In: Blaquirre, A. (Ed.): Topics in Differential Games. North-Holland Publishing Company, Amsterdam, 25-31
57. Karlin, S. (1959): Mathematical Methods and Theory in Game, Programming and Economics, Vol.II, Pergamon Press, London
58. Kikuta, K. (1990): A Hide and Seek Game with Traveling Cost. Journal of the Operations Research Society of Japan 33, 168-187
59. Kikuta, K. (1990): A One Dimensional Search Game with Traveling Cost. Journal of the Operations Research Society of Japan 33, 262-276
60. Kikuta, K. (1991): A Search Game with Traveling Cost. Journal of the Operations Research Society of Japan 34, 365-382
61. Kikuta, K., Ruckle, W.H. (1994): Initial Point Search on Weighted Trees. Naval Research Logistics 41, 821-831
62. Kikuta, K., Ruckle, W.H. (1997): Accumulation Games, Part 1: Noisy Search. Journal of Optimization Theory and Applications 94, 395-408
63. Koopman, B.O. (1956): The Theory of Search. Part II,III. Operations Research 4, 503-531
64. Koopman, B.O. (1980): Search and Screening. Pergamon Press, New York
65. Kurisu, T. (1980): On a Three-Person Silent Marksmanship Contest. Journal of the Operations Research Society of Japan 23, 326-339
66. Kurisu, T. (1989): On a Duel with Time Lag and Equal Accuracy Functions. Journal of Optimazition Theory and Applications 60, 43-56
67. Lalley, S.P. (1988): A One-Dimensional Infiltration Game. Naval Research Logistics 35, 441-446
68. Lalley, S.P., Robbins, H. (1988): Stochastic Search in a Convex Region. Probability Theory and Related Fields 77, 99-116
69. Lee, K.T. (1983): A Firing Game with a Time Lag. Journal of Optimization Theory and Applications 41, 548-547
70. Lee, K.T. (1985): An Evasion with a Destination. Journal of Optimization Theory and Applications 46, 359-372
71. Mangasarian, O.L. (1969): Nonlinear Programming. McGraw-Hill, New York
72. Maynard Smith, J. (1982): Evolution and the Theory of Games. Cambrige University Press, Cambridge
73. Monahan, G.E. (1987): The Structure of Equilibria in Market Share Attraction Models. Management Sciences 33, 228 - 243

74. Monahan, G.E. (1996): Finding Saddle Points on Polyhedra: Solving Certain Continuous Minimax Problems. Naval Research Logistics 43, 821-837
75. Nakai, T. (1986): A Sequential Evasion-Search Games with a Goal. Journal Operations Research of Japan 29, 113-122
76. Nakai, T. (1988): Search Models with Continuous Effort Under Various Criteria. Journal of the Operations Research Society of Japan 31, 335 - 351
77. Owen, G. (1982): Game Theory. W.B.Sanders, Philadelphia
78. Pavlopic, L. (1995): Search for an Infiltrator. Naval Research Logistics 42, 15-26
79. Pavlopic, L. (1995): A Search Game on the Union of Graphs with Immobile Hider. Naval Research Logistics 42: 1177-1189
80. Petrosjan, L.A. (1993): Differential Games of Pursuit. World Scientific, London
81. Petrosjan, L.A., Zenkevich, N.A. (1996): Game Theory. World Scientific, London
82. Rasmusen, E. (1989): Games and Information; An Introduction to Game Theory. Blackwell, Oxford
83. Reijnierse, J.H., Potters, J.M. (1993): Search Games with Immobile Hider. International Journal of Game Theory 21, 385-396
84. Roberts, D., Gittins, J. (1978): The Search for an Intellegent Evader: Strategies for Searcher and Evader in the Two Regions Problem. Naval Research Logistics Quarterly 25, 95-106
85. Ruckle, W.H. (1980): On the Constractability of Solution a Pair of Two Person Search Games. International Journal of Game Theory 8, 235-240
86. Ruckle, W.H. (1981): Ambushing Random Walks II. Operations Research 29, 108-120
87. Ruckle, W.H. (1983): Geometric Games and Their Applications. Pitman Advansed Publishing Program, Boston
88. Ruckle, W.H. (1990): A Discrete game of Infiltration. (unpublised)
89. Ruckle, W.H. (1992): The Upper Risk of an Inspection Agreement. Operations Research 40, 877-894
90. Ruckle, W.H. (1992): The Reporting Time Problem: Integration of Intelligence with Verification. Naval Research Logistics 39, 893-903
91. Sakaguchi, M. (1973): Two-sided Search Games. Journal of the Operations Research Society of Japan 16, 207-225
92. Sakaguchi, M. (1978): Marksmanship contests: Non-zero-sum Game of Timing. Mathematica Japonica 22, 457-469
93. Sakaguchi, M. (1986): A Pursuit Game on a Line with a Bunker for the Evader. Mathematica Japonica 31, 449-461
94. Sakaguchi, M. (1987): Some Simple Models of Duels with Random Termination Time. Mathematica Japonica 32, 833-848
95. Sakaguchi, M. (1987): A Two-sided Resource Allocation Game in Search for a Stationary Object. Mathematica Japonica 32, 979-991
96. Sakaguchi, M. (1988): The Value of Sample Information in High - Hand - Wins Poker. Mathematica Japonica 33, 587-607
97. Sakaguchi, M. (1988): The Value of Sample Information in La Relance Poker. Mathematica Japonica 33, 777-800
98. Sakaguchi, M. (1993): On Two and Three Person Exchange Games. Mathematica Japonica 38, 791-801
99. Sakaguchi, M. (1994): A Sequential Game of Multi-Opportunity Infiltration. Mathematica Japonica 39, 157-166
100. Sakaguchi, M., Mazalov, V.V. (1996): Two-Person Hi-Lo poker - Stud and Draw. Mathematica Japonica 44, 39-53
101. Stone, L.D. (1975): Theory of Optimal Search. Academic Press, New York

102. Subelman, E.J. (1981): A Hide-Search Game. Journal of Applied Probability 18, 628-640
103. Sweat, C.W. (1971): A Single-Shot Noisy Duel with Detection Uncertainty. Operations Research 19, 170-181
104. Teraoka, Y. (1983): A Two Person Game of Timing with Random Termination. Journal Optimization Theory and Applications 40, 379-396
105. Teraoka, Y. (1986): Silent-noisy Marksmanship Contest with Random Termination. Journal of Optimization Theory and Applications 49, 477-487
106. Teraoka, Y., Yamada, Y. (1997): Games of Production Development in Manufacturing. In: Christer, A., Osaki, S., Thomas, L. (Eds.): Stochastic Modelling in Innovative Manufacturing. Springer-Verlag, Berlin Heidelberg New York Tokyo, 58-67
107. Thomas, L.C., Washburn, A.R. (1991): Dynamic Search Games. Operations Research 39, 415-422
108. Thomas, M.U., Nisgav, Y. (1976): An Infiltration Game with Time Dependent Payoff. Naval Research Logistics 23, 297-303
109. Von-Neumann, J. (1928): Zur Theorie Der Gesellschftesspiele. Mathematische Annalen 100, 295-320
110. Washburn, A. (1980): Search Evasion Game in a Fixed Region. Operations Research 28: 1290-1298
111. Worsham, R.H. (1974): A Discrete Game with a Mobile Hider. In: Roxin, E.O., Liu, T.P., Stenberg, R.L. (Eds.): Differential Games and Control Theory, Marcel Dekker, New York, 201-230
112. Zoroa, M., Zoroa, P. (1993): Some Games of Search on a Lattice. Naval Research Logistics 40, 525-541

Lecture Notes in Economics and Mathematical Systems

For information about Vols. 1–295
please contact your bookseller or Springer-Verlag

Vol. 300: U. Derigs, Programming in Networks and Graphs. XI, 315 pages. 1988.

Vol. 301: J. Kacprzyk, M. Roubens (Eds.), Non-Conventional Preference Relations in Decision Making. VII, 155 pages. 1988.

Vol. 302: H.A. Eiselt, G. Pederzoli (Eds.), Advances in Optimization and Control. Proceedings, 1986. VIII, 372 pages. 1988.

Vol. 303: F.X. Diebold, Empirical Modeling of Exchange Rate Dynamics. VII, 143 pages. 1988.

Vol. 304: A. Kurzhanski, K. Neumann, D. Pallaschke (Eds.), Optimization, Parallel Processing and Applications. Proceedings, 1987. VI, 292 pages. 1988.

Vol. 305: G.-J.C.Th. van Schijndel, Dynamic Firm and Investor Behaviour under Progressive Personal Taxation. X, 215 pages.1988.

Vol. 306: Ch. Klein, A Static Microeconomic Model of Pure Competition. VIII, 139 pages. 1988.

Vol. 307: T.K. Dijkstra (Ed.), On Model Uncertainty and its Statistical Implications. VII, 138 pages. 1988.

Vol. 308: J.R. Daduna, A. Wren (Eds.), Computer-Aided Transit Scheduling. VIII, 339 pages. 1988.

Vol. 309: G. Ricci, K. Velupillai (Eds.), Growth Cycles and Multisectoral Economics: The Goodwin Tradition. III, 126 pages. 1988.

Vol. 310: J. Kacprzyk, M. Fedrizzi (Eds.), Combining Fuzzy Imprecision with Probabilistic Uncertainty in Decision Making. IX, 399 pages. 1988.

Vol. 311: R. Färe, Fundamentals of Production Theory. IX, 163 pages. 1988.

Vol. 312: J. Krishnakumar, Estimation of Simultaneous Equation Models with Error Components Structure. X, 357 pages. 1988.

Vol. 313: W. Jammernegg, Sequential Binary Investment Decisions. VI, 156 pages. 1988.

Vol. 314: R. Tietz, W. Albers, R. Selten (Eds.), Bounded Rational Behavior in Experimental Games and Markets. VI, 368 pages. 1988.

Vol. 315: I. Orishimo, G.J.D. Hewings, P. Nijkamp (Eds), Information Technology: Social and Spatial Perspectives. Proceedings 1986. VI, 268 pages. 1988.

Vol. 316: R.L. Basmann, D.J. Slottje, K. Hayes, J.D. Johnson, D.J. Molina, The Generalized Fechner-Thurstone Direct Utility Function and Some of its Uses. VIII, 159 pages. 1988.

Vol. 317: L. Bianco, A. La Bella (Eds.), Freight Transport Planning and Logistics. Proceedings, 1987. X, 568 pages. 1988.

Vol. 318: T. Doup, Simplicial Algorithms on the Simplotope. VIII, 262 pages. 1988.

Vol. 319: D.T. Luc, Theory of Vector Optimization. VIII, 173 pages. 1989.

Vol. 320: D. van der Wijst, Financial Structure in Small Business. VII, 181 pages. 1989.

Vol. 321: M. Di Matteo, R.M. Goodwin, A. Vercelli (Eds.), Technological and Social Factors in Long Term Fluctuations. Proceedings. IX, 442 pages. 1989.

Vol. 322: T. Kollintzas (Ed.), The Rational Expectations Equilibrium Inventory Model. XI, 269 pages. 1989.

Vol. 323: M.B.M. de Koster, Capacity Oriented Analysis and Design of Production Systems. XII, 245 pages. 1989.

Vol. 324: I.M. Bomze, B.M. Pötscher, Game Theoretical Foundations of Evolutionary Stability. VI, 145 pages. 1989.

Vol. 325: P. Ferri, E. Greenberg, The Labor Market and Business Cycle Theories. X, 183 pages. 1989.

Vol. 326: Ch. Sauer, Alternative Theories of Output, Unemployment, and Inflation in Germany: 1960–1985. XIII, 206 pages. 1989.

Vol. 327: M. Tawada, Production Structure and International Trade. V, 132 pages. 1989.

Vol. 328: W. Güth, B. Kalkofen, Unique Solutions for Strategic Games. VII, 200 pages. 1989.

Vol. 329: G. Tillmann, Equity, Incentives, and Taxation. VI, 132 pages. 1989.

Vol. 330: P.M. Kort, Optimal Dynamic Investment Policies of a Value Maximizing Firm. VII, 185 pages. 1989.

Vol. 331: A. Lewandowski, A.P. Wierzbicki (Eds.), Aspiration Based Decision Support Systems. X, 400 pages. 1989.

Vol. 332: T.R. Gulledge, Jr., L.A. Litteral (Eds.), Cost Analysis Applications of Economics and Operations Research. Proceedings. VII, 422 pages. 1989.

Vol. 333: N. Dellaert, Production to Order. VII, 158 pages. 1989.

Vol. 334: H.-W. Lorenz, Nonlinear Dynamical Economics and Chaotic Motion. XI, 248 pages. 1989.

Vol. 335: A.G. Lockett, G. Islei (Eds.), Improving Decision Making in Organisations. Proceedings. IX, 606 pages. 1989.

Vol. 336: T. Puu, Nonlinear Economic Dynamics. VII, 119 pages. 1989.

Vol. 337: A. Lewandowski, I. Stanchev (Eds.), Methodology and Software for Interactive Decision Support. VIII, 309 pages. 1989.

Vol. 338: J.K. Ho, R.P. Sundarraj, DECOMP: An Implementation of Dantzig-Wolfe Decomposition for Linear Programming. VI, 206 pages.

Vol. 339: J. Terceiro Lomba, Estimation of Dynamic Econometric Models with Errors in Variables. VIII, 116 pages. 1990.

Vol. 340: T. Vasko, R. Ayres, L. Fontvieille (Eds.), Life Cycles and Long Waves. XIV, 293 pages. 1990.

Vol. 341: G.R. Uhlich, Descriptive Theories of Bargaining. IX, 165 pages. 1990.

Vol. 342: K. Okuguchi, F. Szidarovszky, The Theory of Oligopoly with Multi-Product Firms. V, 167 pages. 1990.

Vol. 343: C. Chiarella, The Elements of a Nonlinear Theory of Economic Dynamics. IX, 149 pages. 1990.

Vol. 344: K. Neumann, Stochastic Project Networks. XI, 237 pages. 1990.

Vol. 345: A. Cambini, E. Castagnoli, L. Martein, P Mazzoleni, S. Schaible (Eds.), Generalized Convexity and Fractional Programming with Economic Applications. Proceedings, 1988. VII, 361 pages. 1990.

Vol. 346: R. von Randow (Ed.), Integer Programming and Related Areas. A Classified Bibliography 1984–1987. XIII, 514 pages. 1990.

Vol. 347: D. Ríos Insua, Sensitivity Analysis in Multiobjective Decision Making. XI, 193 pages. 1990.

Vol. 348: H. Störmer, Binary Functions and their Applications. VIII, 151 pages. 1990.

Vol. 349: G.A. Pfann, Dynamic Modelling of Stochastic Demand for Manufacturing Employment. VI, 158 pages. 1990.

Vol. 350: W.-B. Zhang, Economic Dynamics. X, 232 pages. 1990.

Vol. 351: A. Lewandowski, V. Volkovich (Eds.), Multiobjective Problems of Mathematical Programming. Proceedings, 1988. VII, 315 pages. 1991.

Vol. 352: O. van Hilten, Optimal Firm Behaviour in the Context of Technological Progress and a Business Cycle. XII, 229 pages. 1991.

Vol. 353: G. Ricci (Ed.), Decision Processes in Economics. Proceedings, 1989. III, 209 pages 1991.

Vol. 354: M. Ivaldi, A Structural Analysis of Expectation Formation. XII, 230 pages. 1991.

Vol. 355: M. Salomon. Deterministic Lotsizing Models for Production Planning. VII, 158 pages. 1991.

Vol. 356: P. Korhonen, A. Lewandowski, J . Wallenius (Eds.), Multiple Criteria Decision Support. Proceedings, 1989. XII, 393 pages. 1991.

Vol. 357: P. Zörnig, Degeneracy Graphs and Simplex Cycling. XV, 194 pages. 1991.

Vol. 358: P. Knottnerus, Linear Models with Correlated Disturbances. VIII, 196 pages. 1991.

Vol. 359: E. de Jong, Exchange Rate Determination and Optimal Economic Policy Under Various Exchange Rate Regimes. VII, 270 pages. 1991.

Vol. 360: P. Stalder, Regime Translations, Spillovers and Buffer Stocks. VI, 193 pages . 1991.

Vol. 361: C. F. Daganzo, Logistics Systems Analysis. X, 321 pages. 1991.

Vol. 362: F. Gehrels, Essays in Macroeconomics of an Open Economy. VII, 183 pages. 1991.

Vol. 363: C. Puppe, Distorted Probabilities and Choice under Risk. VIII, 100 pages . 1991

Vol. 364: B. Horvath, Are Policy Variables Exogenous? XII, 162 pages. 1991.

Vol. 365: G. A. Heuer, U. Leopold-Wildburger. Balanced Silverman Games on General Discrete Sets. V, 140 pages. 1991.

Vol. 366: J. Gruber (Ed.), Econometric Decision Models. Proceedings, 1989. VIII, 636 pages. 1991.

Vol. 367: M. Grauer, D. B. Pressmar (Eds.), Parallel Computing and Mathematical Optimization. Proceedings. V, 208 pages. 1991.

Vol. 368: M. Fedrizzi, J. Kacprzyk, M. Roubens (Eds.), Interactive Fuzzy Optimization. VII, 216 pages. 1991.

Vol. 369: R. Koblo, The Visible Hand. VIII, 131 pages.1991.

Vol. 370: M. J. Beckmann, M. N. Gopalan, R. Subramanian (Eds.), Stochastic Processes and their Applications. Proceedings, 1990. XLI, 292 pages. 1991.

Vol. 371: A. Schmutzler, Flexibility and Adjustment to Information in Sequential Decision Problems. VIII, 198 pages. 1991.

Vol. 372: J. Esteban, The Social Viability of Money. X, 202 pages. 1991.

Vol. 373: A. Billot, Economic Theory of Fuzzy Equilibria. XIII, 164 pages. 1992.

Vol. 374: G. Pflug, U. Dieter (Eds.), Simulation and Optimization. Proceedings, 1990. X, 162 pages. 1992.

Vol. 375: S.-J. Chen, Ch.-L. Hwang, Fuzzy Multiple Attribute Decision Making. XII, 536 pages. 1992.

Vol. 376: K.-H. Jöckel, G. Rothe, W. Sendler (Eds.), Bootstrapping and Related Techniques. Proceedings, 1990. VIII, 247 pages. 1992.

Vol. 377: A. Villar, Operator Theorems with Applications to Distributive Problems and Equilibrium Models. XVI, 160 pages. 1992.

Vol. 378: W. Krabs, J. Zowe (Eds.), Modern Methods of Optimization. Proceedings, 1990. VIII, 348 pages. 1992.

Vol. 379: K. Marti (Ed.), Stochastic Optimization. Proceedings, 1990. VII, 182 pages. 1992.

Vol. 380: J. Odelstad, Invariance and Structural Dependence. XII, 245 pages. 1992.

Vol. 381: C. Giannini, Topics in Structural VAR Econometrics. XI, 131 pages. 1992.

Vol. 382: W. Oettli, D. Pallaschke (Eds.), Advances in Optimization. Proceedings, 1991. X, 527 pages. 1992.

Vol. 383: J. Vartiainen, Capital Accumulation in a Corporatist Economy. VII, 177 pages. 1992.

Vol. 384: A. Martina, Lectures on the Economic Theory of Taxation. XII, 313 pages. 1992.

Vol. 385: J. Gardeazabal, M. Regúlez, The Monetary Model of Exchange Rates and Cointegration. X, 194 pages. 1992.

Vol. 386: M. Desrochers, J.-M. Rousseau (Eds.), Computer-Aided Transit Scheduling. Proceedings, 1990. XIII, 432 pages. 1992.

Vol. 387: W. Gaertner, M. Klemisch-Ahlert, Social Choice and Bargaining Perspectives on Distributive Justice. VIII, 131 pages. 1992.

Vol. 388: D. Bartmann, M. J. Beckmann, Inventory Control. XV, 252 pages. 1992.

Vol. 389: B. Dutta, D. Mookherjee, T. Parthasarathy, T. Raghavan, D. Ray, S. Tijs (Eds.), Game Theory and Economic Applications. Proceedings, 1990. IX, 454 pages. 1992.

Vol. 390: G. Sorger, Minimum Impatience Theorem for Recursive Economic Models. X, 162 pages. 1992.

Vol. 391: C. Keser, Experimental Duopoly Markets with Demand Inertia. X, 150 pages. 1992.

Vol. 392: K. Frauendorfer, Stochastic Two-Stage Programming. VIII, 228 pages. 1992.

Vol. 393: B. Lucke, Price Stabilization on World Agricultural Markets. XI, 274 pages. 1992.

Vol. 394: Y.-J. Lai, C.-L. Hwang, Fuzzy Mathematical Programming. XIII, 301 pages. 1992.

Vol. 395: G. Haag, U. Mueller, K. G. Troitzsch (Eds.), Economic Evolution and Demographic Change. XVI, 409 pages. 1992.

Vol. 396: R. V. V. Vidal (Ed.), Applied Simulated Annealing. VIII, 358 pages. 1992.

Vol. 397: J. Wessels, A. P. Wierzbicki (Eds.), User-Oriented Methodology and Techniques of Decision Analysis and Support. Proceedings, 1991. XII, 295 pages. 1993.

Vol. 398: J.-P. Urbain, Exogeneity in Error Correction Models. XI, 189 pages. 1993.

Vol. 399: F. Gori, L. Geronazzo, M. Galeotti (Eds.), Nonlinear Dynamics in Economics and Social Sciences. Proceedings, 1991. VIII, 367 pages. 1993.

Vol. 400: H. Tanizaki, Nonlinear Filters. XII, 203 pages. 1993.

Vol. 401: K. Mosler, M. Scarsini, Stochastic Orders and Applications. V, 379 pages. 1993.

Vol. 402: A. van den Elzen, Adjustment Processes for Exchange Economies and Noncooperative Games. VII, 146 pages. 1993.

Vol. 403: G. Brennscheidt, Predictive Behavior. VI, 227 pages. 1993.

Vol. 404: Y.-J. Lai, Ch.-L. Hwang, Fuzzy Multiple Objective Decision Making. XIV, 475 pages. 1994.

Vol. 405: S. Komlósi, T. Rapcsák, S. Schaible (Eds.), Generalized Convexity. Proceedings, 1992. VIII, 404 pages. 1994.

Vol. 406: N. M. Hung, N. V. Quyen, Dynamic Timing Decisions Under Uncertainty. X, 194 pages. 1994.

Vol. 407: M. Ooms, Empirical Vector Autoregressive Modeling. XIII, 380 pages. 1994.

Vol. 408: K. Haase, Lotsizing and Scheduling for Production Planning. VIII, 118 pages. 1994.

Vol. 409: A. Sprecher, Resource-Constrained Project Scheduling. XII, 142 pages. 1994.

Vol. 410: R. Winkelmann, Count Data Models. XI, 213 pages. 1994.

Vol. 411: S. Dauzère-Péres, J.-B. Lasserre, An Integrated Approach in Production Planning and Scheduling. XVI, 137 pages. 1994.

Vol. 412: B. Kuon, Two-Person Bargaining Experiments with Incomplete Information. IX, 293 pages. 1994.

Vol. 413: R. Fiorito (Ed.), Inventory, Business Cycles and Monetary Transmission. VI, 287 pages. 1994.

Vol. 414: Y. Crama, A. Oerlemans, F. Spieksma, Production Planning in Automated Manufacturing. X, 210 pages. 1994.

Vol. 415: P. C. Nicola, Imperfect General Equilibrium. XI, 167 pages. 1994.

Vol. 416: H. S. J. Cesar, Control and Game Models of the Greenhouse Effect. XI, 225 pages. 1994.

Vol. 417: B. Ran, D. E. Boyce, Dynamic Urban Transportation Network Models. XV, 391 pages. 1994.

Vol. 418: P. Bogetoft, Non-Cooperative Planning Theory. XI, 309 pages. 1994.

Vol. 419: T. Maruyama, W. Takahashi (Eds.), Nonlinear and Convex Analysis in Economic Theory. VIII, 306 pages. 1995.

Vol. 420: M. Peeters, Time-To-Build. Interrelated Investment and Labour Demand Modelling. With Applications to Six OECD Countries. IX, 204 pages. 1995.

Vol. 421: C. Dang, Triangulations and Simplicial Methods. IX, 196 pages. 1995.

Vol. 422: D. S. Bridges, G. B. Mehta, Representations of Preference Orderings. X, 165 pages. 1995.

Vol. 423: K. Marti, P. Kall (Eds.), Stochastic Programming. Numerical Techniques and Engineering Applications. VIII, 351 pages. 1995.

Vol. 424: G. A. Heuer, U. Leopold-Wildburger, Silverman's Game. X, 283 pages. 1995.

Vol. 425: J. Kohlas, P.-A. Monney, A Mathematical Theory of Hints. XIII, 419 pages, 1995.

Vol. 426: B. Finkenstädt, Nonlinear Dynamics in Economics. IX, 156 pages. 1995.

Vol. 427: F. W. van Tongeren, Microsimulation Modelling of the Corporate Firm. XVII, 275 pages. 1995.

Vol. 428: A. A. Powell, Ch. W. Murphy, Inside a Modern Macroeconometric Model. XVIII, 424 pages. 1995.

Vol. 429: R. Durier, C. Michelot, Recent Developments in Optimization. VIII, 356 pages. 1995.

Vol. 430: J. R. Daduna, I. Branco, J. M. Pinto Paixão (Eds.), Computer-Aided Transit Scheduling. XIV, 374 pages. 1995.

Vol. 431: A. Aulin, Causal and Stochastic Elements in Business Cycles. XI, 116 pages. 1996.

Vol. 432: M. Tamiz (Ed.), Multi-Objective Programming and Goal Programming. VI, 359 pages. 1996.

Vol. 433: J. Menon, Exchange Rates and Prices. XIV, 313 pages. 1996.

Vol. 434: M. W. J. Blok, Dynamic Models of the Firm. VII, 193 pages. 1996.

Vol. 435: L. Chen, Interest Rate Dynamics, Derivatives Pricing, and Risk Management. XII, 149 pages. 1996.

Vol. 436: M. Klemisch-Ahlert, Bargaining in Economic and Ethical Environments. IX, 155 pages. 1996.

Vol. 437: C. Jordan, Batching and Scheduling. IX, 178 pages. 1996.

Vol. 438: A. Villar, General Equilibrium with Increasing Returns. XIII, 164 pages. 1996.

Vol. 439: M. Zenner, Learning to Become Rational. VII, 201 pages. 1996.

Vol. 440: W. Ryll, Litigation and Settlement in a Game with Incomplete Information. VIII, 174 pages. 1996.

Vol. 441: H. Dawid, Adaptive Learning by Genetic Algorithms. IX, 166 pages.1996.

Vol. 442: L. Corchón, Theories of Imperfectly Competitive Markets. XIII, 163 pages. 1996.

Vol. 443: G. Lang, On Overlapping Generations Models with Productive Capital. X, 98 pages. 1996.

Vol. 444: S. Jørgensen, G. Zaccour (Eds.), Dynamic Competitive Analysis in Marketing. X, 285 pages. 1996.

Vol. 445: A. H. Christer, S. Osaki, L. C. Thomas (Eds.), Stochastic Modelling in Innovative Manufactoring. X, 361 pages. 1997.

Vol. 446: G. Dhaene, Encompassing. X, 160 pages. 1997.

Vol. 447: A. Artale, Rings in Auctions. X, 172 pages. 1997.

Vol. 448: G. Fandel, T. Gal (Eds.), Multiple Criteria Decision Making. XII, 678 pages. 1997.

Vol. 449: F. Fang, M. Sanglier (Eds.), Complexity and Self-Organization in Social and Economic Systems. IX, 317 pages, 1997.

Vol. 450: P. M. Pardalos, D. W. Hearn, W. W. Hager, (Eds.), Network Optimization. VIII, 485 pages, 1997.

Vol. 451: M. Salge, Rational Bubbles. Theoretical Basis, Economic Relevance, and Empirical Evidence with a Special Emphasis on the German Stock Market.IX, 265 pages. 1997.

Vol. 452: P. Gritzmann, R. Horst, E. Sachs, R. Tichatschke (Eds.), Recent Advances in Optimization. VIII, 379 pages. 1997.

Vol. 453: A. S. Tangian, J. Gruber (Eds.), Constructing Scalar-Valued Objective Functions. VIII, 298 pages. 1997.

Vol. 454: H.-M. Krolzig, Markov-Switching Vector Autoregressions. XIV, 358 pages. 1997.

Vol. 455: R. Caballero, F. Ruiz, R. E. Steuer (Eds.), Advances in Multiple Objective and Goal Programming. VIII, 391 pages. 1997.

Vol. 456: R. Conte, R. Hegselmann, P. Terna (Eds.), Simulating Social Phenomena. VIII, 536 pages. 1997.

Vol. 457: C. Hsu, Volume and the Nonlinear Dynamics of Stock Returns. VIII, 133 pages. 1998.

Vol. 458: K. Marti, P. Kall (Eds.), Stochastic Programming Methods and Technical Applications. X, 437 pages. 1998.

Vol. 459: H. K. Ryu, D. J. Slottje, Measuring Trends in U.S. Income Inequality. XI, 195 pages. 1998.

Vol. 460: B. Fleischmann, J. A. E. E. van Nunen, M. G. Speranza, P. Stähly, Advances in Distribution Logistic. XI, 535 pages. 1998.

Vol. 461: U. Schmidt, Axiomatic Utility Theory under Risk. XV, 201 pages. 1998.

Vol. 462: L. von Auer, Dynamic Preferences, Choice Mechanisms, and Welfare. XII, 226 pages. 1998.

Vol. 463: G. Abraham-Frois (Ed.), Non-Linear Dynamics and Endogenous Cycles. VI, 204 pages. 1998.

Vol. 464: A. Aulin, The Impact of Science on Economic Growth and its Cycles. IX, 204 pages. 1998.

Vol. 465: T. J. Stewart, R. C. van den Honert (Eds.), Trends in Multicriteria Decision Making. X, 448 pages. 1998.

Vol. 466: A. Sadrieh, The Alternating Double Auction Market. VII, 350 pages. 1998.

Vol. 467: H. Hennig-Schmidt, Bargaining in a Video Experiment. Determinants of Boundedly Rational Behavior. XII, 221 pages. 1999.

Vol. 468: A. Ziegler, A Game Theory Analysis of Options. XIV, 145 pages. 1999.

Vol. 469: M. P. Vogel, Environmental Kuznets Curves. XIII, 197 pages. 1999.

Vol. 470: M. Ammann, Pricing Derivative Credit Risk. XII, 228 pages. 1999.

Vol. 471: N. H. M. Wilson (Ed.), Computer-Aided Transit Scheduling. XI, 444 pages. 1999.

Vol. 472: J.-R. Tyran, Money Illusion and Strategic Complementarity as Causes of Monetary Non-Neutrality. X, 228 pages. 1999.

Vol. 473: S. Helber, Performance Analysis of Flow Lines with Non-Linear Flow of Material. IX, 280 pages. 1999.

Vol. 474: U. Schwalbe, The Core of Economies with Asymmetric Information. IX, 141 pages. 1999.

Vol. 475: L. Kaas, Dynamic Macroelectronics with Imperfect Competition. XI, 155 pages. 1999.

Vol. 476: R. Demel, Fiscal Policy, Public Debt and the Term Structure of Interest Rates. X, 279 pages. 1999.

Vol. 477: M. Théra, R. Tichatschke (Eds.), Ill-posed Variational Problems and Regularization Techniques. VIII, 274 pages. 1999.

Vol. 478: S. Hartmann, Project Scheduling under Limited Resources. XII, 221 pages. 1999.

Vol. 479: L. v. Thadden, Money, Inflation, and Capital Formation. IX, 192 pages. 1999.

Vol. 480: M. Grazia Speranza, P. Stähly (Eds.), New Trends in Distribution Logistics. X, 336 pages. 1999.

Vol. 481: V. H. Nguyen, J. J. Strodiot, P. Tossings (Eds.). Optimations. IX, 498 pages. 2000.

Vol. 482: W. B. Zhang, A Theory of International Trade. XI, 192 pages. 2000.

Vol. 483: M. Königstein, Equity, Efficiency and Evolutionary Stability in Bargaining Games with Joint Production. XII, 197 pages. 2000.

Vol. 484: D. D. Gatti, M. Gallegati, A. Kirman, Interaction and Market Structure. VI, 298 pages. 2000.

Vol. 485: A. Garnaev, Search Games and Other Applications of Game Theory. VIII, 145 pages. 2000.

Vol. 486: M. Neugart, Nonlinear Labor Market Dynamics. X, 175 pages. 2000.